BAHAMAS PRIMARY

Mathematics Book 5

The authors and publishers would like to thank the following members of the Teachers' Panel, who have assisted in the planning, content and development of the books:

Chairperson: Dr Joan Rolle, Senior Education Officer, Primary School Mathematics, Department of Education

Team members:

Lelani Burrows, Anglican Education Authority

Deidre Cooper, Catholic Board of Education

LeAnna T. Deveaux-Miller, T.G. Glover Professional Development and Research School

Dr Marcella Elliott Ferguson, University of The Bahamas

Theresa McPhee, Education Officer, High School Mathematics, Department of Education

Joelynn Stubbs, C.W. Sawyer Primary School

Dyontaleé Turnquest Rolle, Eva Hilton Primary School

Karen Morrison, Rentia Pretorius and Lisa Greenstein

HODDER EDUCATION
AN HACHETTE UK COMPANY

The Publishers would like to thank the following for permission to reproduce copyright material.

Photo credits

All photos © Mike van der Wolk, Tel: +27 83 2686000, mike@springhigh.co.za, except: page 127 © Richard Ellis/Alamy Stock Photo.

Hachette UK's policy is to use papers that are natural, renewable and recyclable products and made from wood grown in sustainable forests. The logging and manufacturing processes are expected to conform to the environmental regulations of the country of origin.

Orders: please contact Bookpoint Ltd, 130 Park Drive, Milton Park, Abingdon, Oxon OX14 4SE. Telephone: (44) 01235 827720. Fax: (44) 01235 400454. Email education@bookpoint.co.uk Lines are open from 9 a.m. to 5 p.m., Monday to Saturday, with a 24-hour message answering service. You can also order through our website: www.hoddereducation.com

ISBN: **978 1 4718 6469 8**

© Cloud Publishing Services 2017

First published in 2017 by
Hodder Education,
An Hachette UK Company
Carmelite House
50 Victoria Embankment
London EC4Y 0DZ

www.hoddereducation.com

Impression number   10 9 8 7 6 5 4 3 2

Year   2021 2020 2019

Cover photo © Shutterstock/Worachat Sodsri

Illustrations by Peter Lubach and Aptara Inc.

Typeset in India by Aptara Inc.

Printed in India

A catalogue record for this title is available from the British Library.

Contents

Topic 1 Getting Ready  Workbook pages 1–3

Key Words

counting
numbers
fractions
decimals
patterns
calculations
shapes
measurement
graphs

▲ Toniqua and her family saw huge numbers of spring flowers in a national park when they visited the USA. How would you estimate the number of flowers in the photograph? This is just a small section of the park. If there are about 175 flowers in a square metre, about how many would there be in an area of 10 000 square metres?

If you think about it, you will realize that you know lots of mathematics already. You are able to **count**, work with large **numbers**, **fractions** and **decimals**, recognize and work with **patterns**, **measure** different quantities, calculate in different ways, draw and make sense of **graphs** and charts, recognize **shapes** and their properties, describe probabilities and use a number of strategies to solve problems.

This year, you will build on what you know and apply the skills you already have to learn more and become even better at mathematics.

Getting Started

1 Tell your partner three things you remember from mathematics last year.

2 Work in pairs to explore what you are going to learn this year. Turn to the contents page of this book. In which topic/s will you work with each of these?

a Kilograms and grams.

b Symmetry and transformations.

c Fact families.

d Divisors and dividends.

e Square units of measurement.

Unit 1 Check Your Skills

Let's Think ...

The numbers in each row, column and diagonal all have the same sum.
This is a magic square.

4	9	2
3	5	7
8	1	6

Is this a magic square? Give a reason for your answer.

5	4	7
9	1	6
2	8	2

Use what you know about counting, numbers, fractions, decimals, patterns, calculations, shapes, measurement and graphs as you work through these activities. Ask for help if you cannot remember what to do.

1 Joshua and Jerome each bought a sweet costing 30¢. They each paid with a $1.00 bill.
   a How much change did they each get?
   b Joshua received two coins as his change. What could they be?
   c Jerome received four coins. What could they be?

2 How many minutes are there in:
   a $1\frac{1}{2}$ hours
   b $2\frac{1}{4}$ hours
   c $3\frac{3}{5}$ hours?

3 How many 200 mL containers can you fill from a drum containing 3 litres of oil?

4 Write each number in expanded notation.
   a 876
   b 2346
   c 80976
   d 123098
   e 1234000

5 What is the value of the digit 8 in each of these numbers?
   a 48976
   b 54008
   c 187300
   d 8999999

6 Work out the answers to these calculations. Use any method you like, but show your working.
   a 12345 + 2765
   b 19876 − 4859
   c 23 × 15
   d 3145 ÷ 5
   e 3456 + 349 + 12807
   f 9 × 387

7 Work out the patterns. Write the next three numbers in each.
   a 1   8   15   22   ☐ ☐ ☐

   b 89   83   77   71   ☐ ☐ ☐

   c 45   450   4500   ☐ ☐ ☐

   d 448   224   112   ☐ ☐ ☐

8 Arrange these periods of time from the longest to the shortest.

> day   decade   hour   leap year   minute   month   second   week   year

9 What do you call a polygon with four equal sides and four equal angles?

10 This year the mail ship sailed 85 604 miles. This is 17 342 miles more than last year. How many miles did it sail last year?

11 Zara has three pieces of ribbon. They are 25 cm, 347 mm and 64 cm long.
   a How many millimetres of ribbon is this?
   b Write the total length of the ribbon in metres using a decimal fraction.

12 Calculate:
   a $\frac{1}{6} + \frac{3}{6}$
   b $\frac{4}{7} + \frac{2}{7} - \frac{3}{7}$
   c $3\frac{1}{5} + 2\frac{3}{5}$
   d $3\frac{8}{9} - 1\frac{5}{9}$

13 Choose the most accurate temperature in each case.
   a Your normal body temperature: 12 °C   36 °C   56 °C
   b Iced water taken from a fridge: 3 °C   28 °C   43 °C
   c Boiling water in a kettle: 56 °C   98 °C   180 °C

14 a Draw a rectangle with sides of 3 cm and $6\frac{1}{2}$ cm.
   b Calculate the perimeter of the rectangle.

15 Kaylee did the following working to solve a word problem. Write a possible problem for this working.

$$3 \times 15 = 3 + 10 + 3 \times 5$$
$$= 30 + 15$$
$$= 45$$

$$\begin{array}{r} {}^{4}1\!\!\!\!/\,{}^{1}\!\!\!\!/0 \\ -\ 45 \\ \hline 105 \end{array}$$

16 What transformation has been done to each object?

b

b

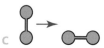
c

17 Say whether each event is impossible, likely or certain.
   a Your cat will do your mathematics homework today.
   b Someone in your class will have a birthday this month.
   c This year, Christmas will fall on the 25th of December.

Looking Back

A rectangle has a perimeter of 24 centimetres. The length is double the width. What is the width of the rectangle?

Topic Review

What Did You Learn?

In this topic, you revised some of the things you learned last year.

Talking Mathematics

Work in pairs to try to answer these children's questions.

> What is the difference between area and volume?

> What words in a problem tell me I need to divide to find the answer?

> James says a rectangle is the same as a square. Why is he incorrect?

> What is a mixed number and how is it different from a whole number?

Quick Check

1 Write the answers only.

   a $27 + 38$
   b $250 + 10$
   c 90×6
   d $400 \div 2$
   e double 120
   f 5×50
   g $19 + 230$
   h $126 - 90$

2 Estimate by rounding and then calculate the answers.

   a $2\,341 + 1\,408$
   b $2\,000 - 1\,209$
   c $1\,987 - 1\,235$
   d $2\,872 - 1\,458$

3 What is the value of each digit in the number $1\,209\,089$?

4 These clocks all show a.m. times. Write the times in order from earliest to latest.

a  b  c  d  e

5 Draw a square inside a right-angled triangle.

6 a Halve each of these digits: 4 6 2 0

   b What is the greatest four-digit number you can write with the new digits?

   c What is the largest possible decimal fraction you can make with 0 in the ones place?

7 The Johnson family has a rectangular vegetable patch that is 5 m long and 4 m wide. They want to put a small fence around the vegetable patch. They will put one fence post every metre. How many posts will they need?

Topic 2 Numbers and Place Value

Workbook pages 4–5

Key Words
number
digit
value
place
place value
millions

▲ In 2015, one port in Florida, USA, imported 27 851 containers of bananas. The bananas weighed 286 902 tons and were worth $82 645 539. Can you say each of these numbers? Which number has 8 in the ten thousands place? What is the value of the 8 in the other two numbers?

Do you remember all the **place values** to **millions**? Last year you worked with large numbers. You counted, ordered and compared numbers to millions and you learned about the **value** of each **place** in the number. This year, you are going to continue to work with large numbers and also extend the place value table to ten millions and hundred millions.

Getting Started

1 Can you write these numbers in numerals?
   a seven million eight hundred twenty-three thousand five hundred twelve
   b twenty-five million

2 What is the number?
   a 1 more than 999
   b 1 more than 9 999
   c 1 more than 9 999 999

3 Say each of these numbers aloud.
   a 8 765 000
   b 3 000 000
   c 340 987
   d 2 300 500
   e 1 450 231

Unit 1 Revisiting Millions

Let's Think …

Torianne has written these digits on the board:

5   7   9   1   2   0   4

a What is the biggest number she can make using them all?

b What is the smallest number she can make using them all?

c Read the numbers.

You know that each digit in a number has a value. In the number 1 265 402, the 1 has a value of 1 million. In the number 2 404 213, the 1 has a value of ten. The place in which the digit is found determines its value.

The place value table shows the places from ones to millions.

Millions	Hundred Thousands	Ten Thousands	Thousands	Hundreds	Tens	Ones
4	2	1	7	4	3	2

To read the number:

- *read the millions first*   *four million*
- *read the thousands next*   *two hundred seventeen thousand*
- *read the rest of the number*   *four hundred thirty-two*

This number is four million two hundred seventeen thousand four hundred thirty-two.

$4\,217\,432 = 4\,000\,000 + 200\,000 + 10\,000 + 7\,000 + 400 + 30 + 2$

Remember to leave a space between millions, thousands and hundreds when you write numbers.

1 Work with a partner. Count from:

    a 99 987 to 1 000 005 by ones

    b 990 000 to 1 001 000 by thousands

    c 999 950 to 1 000 020 by tens

    d 600 000 to 1 200 000 by hundred thousands

    e 1 000 010 back to 999 950 by tens

    f 1 000 002 back to 999 990 by twos.

2 Write the value of the blue digit in each number.

    a 3 243 809   b 4 530 987   c 3 198 087   d 1 209 999

    e 1 098 765   f 987 912   g 1 345 983   h 9 129 999

3 Write these numbers in ascending order.

    6 897   123   1 096 321   900 324   2 000 000   78 098   993 098

4 Ms Newton wrote some numbers on the board.

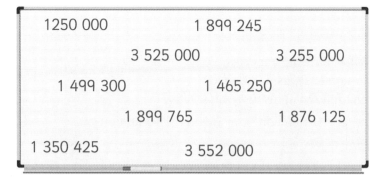

1250 000	1 899 245
3 525 000	3 255 000
1 499 300	1 465 250
1 899 765	1 876 125
1 350 425	3 552 000

a Four students each chose a number from the board. Which number did each student choose?

My number has 5 hundred thousands and 5 ten thousands.

Mine has more than 2 tens and 6 ten thousands.

I have got a number that is greater than three million five hundred thousand with a 2 in the thousands place.

My number is less than 1 500 000 and it has 3 hundreds.

b Ms Newton drew this diagram to sort the numbers.

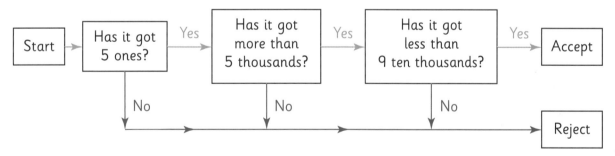

Start → Has it got 5 ones? → Yes → Has it got more than 5 thousands? → Yes → Has it got less than 9 ten thousands? → Yes → Accept

Has it got 5 ones? → No
Has it got more than 5 thousands? → No
Has it got less than 9 ten thousands? → No
→ Reject

Which of the numbers on the board will be accepted by the machine?

c Design your own number sorting diagram that will only accept the number 3 255 000. You must have three sorting questions.

Looking Back

1 Write in numerals, the number that is:
   a 1 million less than 2 500 000
   b 55 000 more than 2 411 000
   c 200 000 more than 3 million
   d 500 000 less than 2 600 000

2 Arrange the numbers you have written in descending order.

Unit 2 More Millions

Let's Think …

One million is written in numerals as 1 000 000.
How would you write the following in numerals? Why?

a 56 million   b 109 million   c 140 million   d $16\frac{1}{2}$ million

The thousands group in the place value table has a place for thousands, ten thousands and hundred thousands.

We can extend the table to the left to make places for ten millions and hundred millions.

Hundred Millions	Ten Millions	Millions	Hundred Thousands	Ten Thousands	Thousands	Hundreds	Tens	Ones
1	5	4	2	1	7	4	3	2

To read the number:

- read the millions first   *one hundred fifty-four million*
- read the thousands next   *two hundred seventeen thousand*
- read the rest of the number *four hundred thirty-two*

This number is one hundred fifty-four million two hundred seventeen thousand four hundred thirty-two.

1 Say each number aloud.
   a 334 000 000   b 210 654 763   c 987 000 987
   d 17 234 900   e 999 435 004   f 599 001 001

2 What is the value of the digit 6 in each number?
   a 564 909 754   b 639 011 400   c 356 999 342

3 Write the greater number in each pair in expanded notation.
   a 645 234 756   647 234 756
   b 812 0135 781   702 144 987
   c 734 680 129   437 860 126

4 Write each set of numbers in ascending order.
   a 12 983 467   10 203 004   8 747 543   24 302 065   18 765 100
   b 199 046 871   132 098 999   191 098 000   218 021 098   200 987 456
   c 231 432 654   231 876 132   213 312 456   312 342 125   231 987 098

5 The diagram shows five planets in our solar system. The table shows the average distance each planet is from the Sun.

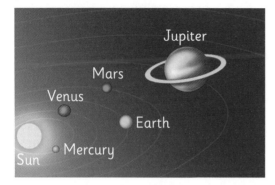

Planet	Average Distance from the Sun (km)
Mercury	57 910 000
Venus	108 200 000
Earth	149 600 000
Mars	227 940 000
Jupiter	788 330 000

a Which of these planets are less than 150 million kilometres from the Sun?

b Which planet is almost 800 million kilometres from the Sun?

c Sharyn says Earth is more than twice as far from the Sun as Mercury. Is she correct?

d Jerome says Earth is about 150 million kilometres from the Sun. Is he correct?

6 The area of each continent is given below in square kilometres.

Continent	Area (square km)
Africa	30 065 000
Antarctica	13 209 000
Asia	44 579 000
Australia	7 687 000
Europe	9 938 000
North America	24 256 000
South America	17 817 000

a Which continents have an area of less than 15 million square kilometres?

b Which continent has the greatest area?

c Which is larger, North America or South America?

d Is Africa bigger or smaller than North America?

e Write the names of the continents in order from smallest to greatest area.

Looking Back

The Atlantic Ocean covers an area of 106 400 000 square kilometres. The Pacific Ocean covers an area of 165 250 000 square kilometres.

a Compare these two areas using a < or > sign.

b What is the value of the digit 6 in each area?

c Approximately how much bigger than the Atlantic is the Pacific Ocean?

Topic Review

What Did You Learn?

- You can write any number using the digits from 0 to 9 and place value.
- The place value table extends to the left as numbers get larger.
- You worked with place value to hundred millions.

Hundred Millions	Ten Millions	Millions	Hundred Thousands	Ten Thousands	Thousands	Hundreds	Tens	Ones

- You read the numbers from left to right in groups.
- When you write the numbers you leave a space between each group of three digits.

Talking Mathematics

Explain in your own words how to:

a say the number 233 876 016

b compare the numbers 123 456 089 and 123 465 908

c work out whether a number is in the hundred millions or not.

Quick Check

1 Write the number you would get if you added:

   a 3 millions to 14 499   b a million to 324 125 000   c 1 to 9 999 999

2 Write each of these numbers in expanded notation.

3 What is the value of the blue digit in each of these numbers?

   a 33 123 457   b 38 765 432   c 309 876 309   d 342 309 765

4 Use all the digits on the card to make:

   a the greatest possible number

   b the least possible number

   c the largest number with 0 in the millions place

   d the smallest number with 3 in the ten millions place.

   4   4   2   3
     3
        0   7
   1   9

5 The area of the oceans is given in the table.

Ocean	Arctic	Atlantic	Indian	Pacific	Southern
Area (sq km)	13 990 000	106 400 000	73 560 000	165 250 000	20 330 000

   a Which ocean is the biggest?

   b Which ocean has an area of less than 20 million square kilometres?

   c Which ocean has an area of more than 100 million square kilometres, but less than 110 million square kilometres?

Topic 3 Exploring Patterns Workbook pages 6–8

▲ Look at these patterned socks. How does each pattern work?
Choose one sock and describe the pattern on it to your partner.
Let them guess which sock you are describing.

You can find number and shape **patterns** everywhere you look. In this topic, you are going to focus on the patterns made by special types of numbers and link these to shape patterns. Patterns are important in mathematics. You are going to investigate some patterns and **describe** how they work.

Getting Started

1 Where can you find the following patterns in your local environment?
  a A pattern of odd and even numbers.
  b A pattern made of repeating rectangular shapes.
  c A pattern containing circles.
  d A pattern made with triangles.
  e A striped pattern.

2 These two number patterns each follow a different rule. Write the rule and work out the next three numbers in each pattern.

  a 13, 15, 17, 19, ☐, ☐, ☐   b 12, 24, 48, 96, ☐, ☐, ☐

3 Look at these two pattern machines. Work out the missing part of the rule and any missing numbers.

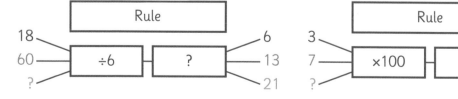

Unit 1 Understand and Describe Patterns

Let's Think …

Look at this pattern.

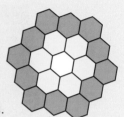

a How many hexagons are there in the centre?

b How many yellow ones are in the first ring?

c How many orange ones are in the second ring?

d Work out how many hexagons you would need to build the next two rings.

e Where would you find a pattern like this in nature?

Patterns can be numerical or geometric.

- *A numerical pattern is a group of numbers that follow each other in a particular order. The rule for the pattern allows you to work out what the numbers are.*

- *A geometric pattern is made using repeated arrangements of lines or shapes. Many geometric patterns can be described using numbers.*

Look at this example carefully to see how geometric and numerical patterns can be linked.

This is a geometric pattern made with rods.

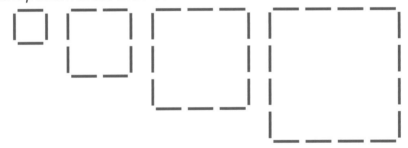

You can describe this pattern in words like this:

- *Each shape in the pattern is a square. Each square is bigger than the one before it. The sides of the squares get longer each time. You add one more rod to the side each time.*

You can also describe the pattern using numbers.

If you list the number of rods in order you get a number sequence: 4, 8, 12, 16, …

You can see from the number sequence that each term is 4 more than the one before it. So, you can add 4 to find the next number. 16 + 4 = 20.

Number sequences are useful for finding the next few numbers in a pattern, but sometimes you need to work out a number much further along the sequence; for example, how many rods will you need to build the 20th shape and the 100th shape in this pattern?

You can use a table to help you work this out.

Shape Number	1	2	3	4	5	20	100
Number of Rods	4	8	12	16			

The table helps you write a rule to work out how many rods you need to build any shape in the pattern.

The rule is: The shape number × 4 = the number of rods

So for the 5th shape   5 × 4 = 20 rods

For the 20th shape   20 × 4 = 80 rods

For the 100th shape   100 × 4 = 400 rods

1 For each pattern:
- draw the next two shapes
- write the pattern as a number sequence
- write a rule for the pattern.

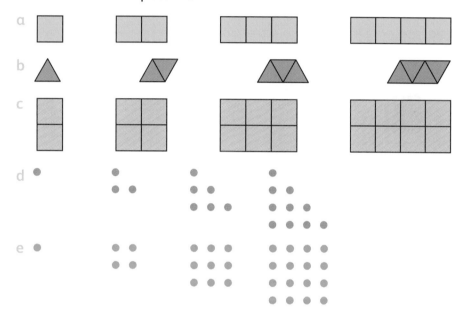

2 Make up a geometric pattern of your own. Exchange patterns with a partner. Try to find the rule used to make your partner's pattern and use it to draw the next two shapes in the pattern.

Looking Back

Look at this pattern.

What will the next shape look like? Why?

Unit 2 Investigate Patterns

Let's Think ...

A class is looking at the dot pattern on the right. The teacher asks how many dots will be needed for the tenth shape. Janine says this pattern is made by adding odd numbers. For the first shape you add one odd number, for the second you add the first two odd numbers and for the third you add the first three odd numbers. So, to find the number of dots in the tenth shape she would add the first ten odd numbers. Maria says there is an easier way to work out the answer because these are square numbers. What does Maria mean?

1   1 + 3 = 4   1 + 3 + 5 = 9

Some numbers make special patterns that you should be able to recognize easily.

Square Numbers

This dot pattern shows the first five square numbers.

1   4   9   16   25

A square number is the result of multiplying a number by itself.

So the fifth square number is $5 \times 5 = 25$ and the tenth square number is $10 \times 10 = 100$.

Triangular Numbers

A triangular number is one that can be shown as a triangular array.

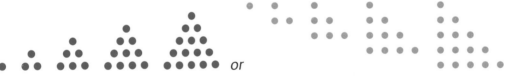 or

These dot patterns show the first five triangular numbers in two different arrangements. You will investigate how to work these numbers out later in this unit.

Powers of Ten

This sequence of numbers shows powers of ten.

10   100   1 000   10 000   100 000   1 000 000

In mathematics, 'to the power of ...' tells you how many times you multiply a number by itself; for example:

10	Ten to the power of 1
10×10	Ten to the power of 2
$10 \times 10 \times 10$	Ten to the power of 3
$10 \times 10 \times 10 \times 10$	Ten to the power of 4

You can work out powers of any number. Square numbers are numbers to the power of 2. So 5×5 is 5 to the power of 2 and 3×3 is 3 to the power of 2.

1 Work out all the square numbers between 0 and 101. List them on a separate piece of paper.

2 Michael builds little houses using dominoes like this:

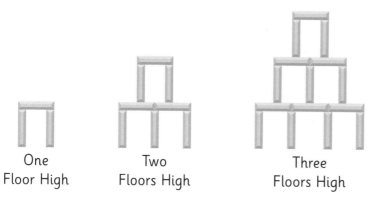

One
Floor High

Two
Floors High

Three
Floors High

   a Without drawing or building the houses, work out how many dominoes Michael needs to build each of the next five houses.

   b List the numbers of dominoes needed for each house. Compare these with the list of square numbers. What do you notice?

3 The first five triangular numbers are 1, 3, 6, 10, 15 …

   a Look at the number sequence. How many are you adding on each time?

   b How does this help you work out the next number in the sequence?

   c Use your method to find all the triangular numbers up to 100.

4 Torianne is trying to find a rule for working out triangular numbers. Her elder sister tells her to think of a triangle as half a rectangle and draws these shapes to show what she means.

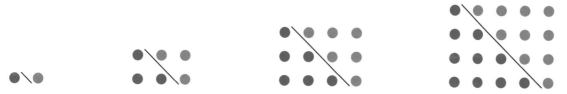

   a Using the number of dots per side as the length, work out the area of each rectangle.

   b Compare that with the triangular numbers.

   c How can you use this information to work out the 9th triangular number without drawing it?

Looking Back

| 9 | 8 | 1 | 6 | 3 | 13 | 2 | 16 | 10 |

Which numbers in the box are:

a square numbers?

b triangular numbers?

c powers of ten?

Topic Review

What Did You Learn?

- A pattern is an ordered set of numbers or objects.
- A geometric pattern is made with shapes or objects.
- A numerical pattern is a sequence of numbers that follow a rule.
- The rule for the pattern can help you extend the pattern and work out missing terms.
- A number multiplied by itself produces a square number.
- Numbers that can be shown as triangular arrays are called triangular numbers.
- A power tells you how many times a number is multiplied by itself. 10 to the power of 3 is $10 \times 10 \times 10$.

Talking Mathematics

The clues for this word puzzle are missing.

Make up a clue for each word.

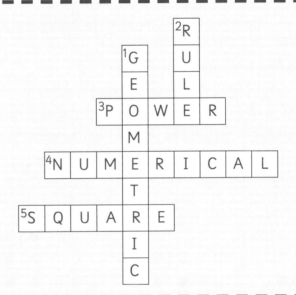

Quick Check

1 Describe each pattern. Write down the next three numbers in the pattern.

    a 1, 4, 7, 10, …
    b 101, 99, 96, 92, …
    c 2 700 000, 270 000, 27 000, …
    d 2 916, 972, 324, …

2 Look at this pattern.

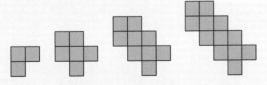

    a Describe in words how it works.
    b How many squares will you need to make the next shape?

    c Write a number sequence to match the pattern.
    d Work out how many squares you would need for the 20th shape in this pattern. Show how you do this.

Topic 4 Temperature Workbook pages 9–12

▲ Can you think of different appliances that we use in our everyday lives to make things hotter or colder? What other ways can we heat or cool things around us without using electrical appliances?

Temperature describes how hot or cold something is. Every day, we notice whether it is warm or cold outside. We dress in order to feel warmer or cooler. We keep food in the refrigerator or freezer to make it last longer. We cook food at high **temperatures**. When we get sick, our bodies get hotter to fight the infection. We may need medicine in order to bring our temperature back down to a healthy level.

Getting Started

1 What are the hottest things you can think of? What are the coldest things you can think of?

2 What instrument do you use to measure temperature?

3 What units do you use to measure temperature?

4 Think of some different temperatures you measured in Grade 4. Match each item to its temperature from the box. Write your answers in a table.

100 °C   60 °C   25 °C   37 °C   0 °C

a Healthy body temperature.

b Boiling point of water.

c Freezing point of water.

d Temperature in The Bahamas in winter.

e Temperature of a cup of tea.

Unit 1 Working with Temperature

Let's Think …

Look at the thermometer in the picture.

1 Which units does it use?

2 What temperature does it show?

3 What are the hottest and coldest temperature readings this thermometer can give?

4 If the temperature is 30 °C and it drops to 25 °C, how many degrees has it dropped by?

5 If the temperature is 12 °C and it goes up by 23 degrees, what is the new temperature?

Temperature Readings

There are two scales of measurement for temperature: Celsius (°C) and Fahrenheit (°F).
In The Bahamas, we mostly use °F. The thermometer above shows 0 °C, which is the same
as 32 °F.

1 Copy the table below. Using the descriptions, write the letter in the Thermometer column that best matches the readings on each thermometer at the top of page 19.

Description	°C	°F	Thermometer
Oven temperature for baking	180	356	
Boiling point for water	100	212	
Cup of tea	60	140	
Hot bath	40	104	
Healthy body temperature	37	98.6	
Warm summer day in the Caribbean	28	82.4	
Freezing point of water	0	32	

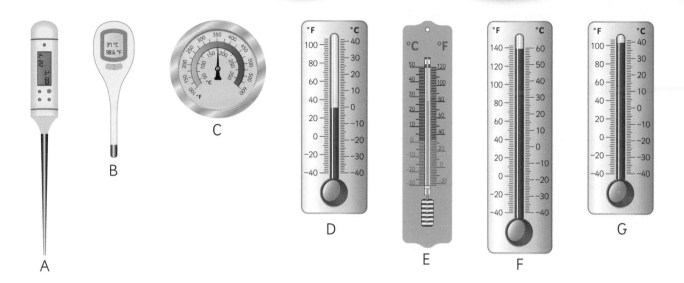

Converting between Celsius and Fahrenheit

Sometimes you need to convert between Celsius and Fahrenheit. Use these rules:

°F to °C   Subtract 32, multiply by 5, then divide by 9.

°C to °F   Multiply by 9, divide by 5, then add 32.

2   Convert the following temperatures.

a   10°C = _____ °F

b   40°F = _____ °C

c   50°C = _____ °F

d   50°F = _____ °C

e   125°C = _____ °F

f   425°F = _____ °C

Working with Negative Temperatures

You know that 0°C is the freezing point of water. Some temperatures are less than zero. The freezer compartment in a refrigerator is usually between −3 and −5°C. At the South Pole, the temperature can be as low as −40°C.

−1°C is one degree less than zero, or 1 degree below zero. −10°C is ten degrees below zero. Remember, when you work with negative numbers, the greater the number after the minus sign, the colder the temperature.

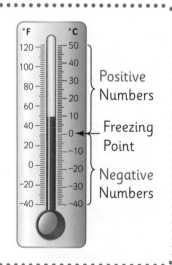

3 Use °F. Write the temperature that is:

   a 5 degrees below zero

   b 23 degrees below zero

   c 50 degrees below zero.

Look at this example.

Katy makes ice cream. The temperature of her mixture is 19 °C. She puts it in the freezer overnight, and the next day the ice cream is frozen and measures −5 °C. What is the change in temperature?

   Step 1 Find the change in temperature from 19 °C to 0 °C.

   The change is 19 °C

   Step 2 Find the change in temperature from 0 °C to −5 °C.

   The change is 5 °C

   Step 3 Add the two changes.

   19 °C + 5 °C = 24 °C

The temperature of the ice cream has changed by 24 °C.

---

4 Use the steps above to solve these problems.

   a John makes ice. The water is 23 °C when he pours it into the trays for freezing. The next day, the ice is −10 °C. What is the temperature change?

   b On a winter's day in Holland, the temperature is 7 °C during the day. It drops to −12 °C at night. What is the change?

   c In Lhasa, Tibet, the temperature can be 29 °C in summer, and it can fall as low as −16 °C in winter. Calculate the change between the summer high and the winter low.

Looking Back

1 Which scale do we use for measuring temperature in The Bahamas?

2 What is the name of the scale more commonly used in the rest of the world?

3 What do we call temperatures below zero?

4 Use the internet or newspapers. Record the temperatures of two islands for one week. Display your information on a line graph or table.

Topic Review

What Did You Learn?

- Temperature tells you how hot or cold something is.
- You measure temperature in degrees Fahrenheit (°F) or degrees Celsius (°C).
- You convert from °F to °C like this: Subtract 32, multiply by 5, then divide by 9.
- You convert from °C to °F like this: Multiply by 9, divide by 5, then add 32.
- Temperatures below zero are negative temperatures.
- The higher the number after the minus sign (−), the lower the temperature.
- To measure the change between a positive and negative temperature, find the difference between the positive temperature and zero; then find the difference between the negative temperature and zero; then add the changes.

Talking Mathematics

- What do you call temperatures below zero?
- What are some of the everyday words you use to describe temperature?

Quick Check

1 Which unit of measure do we use to measure temperature in The Bahamas?

2 Give any two typical temperatures in °C or °F (for example body temperature, or the freezing or boiling point of water).

3 Which temperature is colder: 0 °C or 0 °F?

4 Sugar melts at about 320 °F. What is this temperature in °C?

5 A scientist makes a solution that measures 55 °C. Then she lowers the temperature to −55 °C. What is the temperature change?

Topic 5 Fractions Workbook pages 13–15

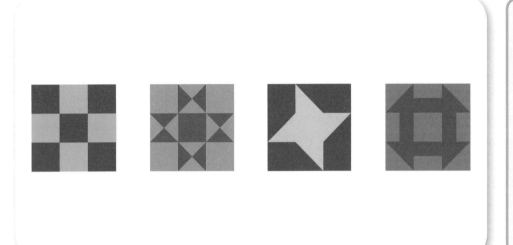

Workbook pages 13–15

Key Words

fraction
part
whole
numerator
denominator
equivalent
simplify
mixed number
order
compare

▲ Mrs Ferguson is making a quilt by joining squares. These are some of the patterns she uses for the squares. Square one is $\frac{4}{9}$ green and $\frac{5}{9}$ purple. What fraction of a square is each colour on the other quilts?

You have already worked with **fractions** such as $\frac{1}{2}, \frac{3}{4}$ and $\frac{9}{10}$ and **mixed numbers** such as $2\frac{1}{2}, 3\frac{1}{5}$ and $2\frac{8}{9}$.

> *Do you remember what the numerator and denominator of a fraction tell you?*
> *The fraction $\frac{3}{4}$ means 3 out of 4 equal parts. The numerator is 3 and the denominator is 4.*

In this topic, you are going to revise basic fractions, work with equivalent fractions and use common factors to simplify fractions. You are also going to **order** and **compare** fractions.

Getting Started

1 Draw a patchwork square design using plain shapes. No colour should occupy more than half of the square.

2 Which of these fractions are greater than $\frac{1}{2}$?

   a $\frac{5}{10}$ b $\frac{7}{8}$ c $\frac{3}{4}$ d $\frac{1}{4}$ e $\frac{3}{7}$ f $\frac{2}{3}$

3 Look at this tray of chocolates:

   a How many chocolates are in half the tray?

   b How many chocolates are left if $\frac{3}{4}$ are eaten?

   c What fraction of the chocolates are shaped like squares?

Unit 1 Revisiting Fractions

Let's Think …

The square has been divided into four parts.

Explain why each part is not $\frac{1}{4}$ of the square.

How would you describe the different parts using fractions?

How much do you remember about fractions? Read through the information in the table below to refresh your memory.

Name	Description	Examples
Fraction	Part of a whole object or set.	One half, three quarters, $\frac{2}{7}$, $\frac{17}{100}$
Numerator	The top number in a fraction. It tells you how many parts out of the whole you are dealing with.	In $\frac{2}{7}$, the numerator is 2. This fraction is 2 out of 7 equal parts.
Denominator	The bottom number in a fraction. It tells you how many equal parts the whole is divided into.	In $\frac{3}{7}$, the denominator is 7. This tells you that the whole is divided into 7 equal parts.
Mixed number	A combination of a whole number and a fraction.	$1\frac{1}{2}$, $4\frac{2}{5}$ These mixed numbers mean you have 1 whole and half of the same sized whole, and 4 wholes and $\frac{2}{5}$ of the same sized whole.
Equivalent fractions	Fractions which have the same value. You can make equivalent fractions by multiplying or dividing the numerator and denominator by the same number.	$\frac{1}{2} = \frac{3}{6}$ $\frac{3}{4} = \frac{60}{40} = \frac{9}{12}$ $\overset{\times 3}{\frac{3}{7}} = \frac{9}{21} \underset{\times 3}{}$ $\overset{\div 4}{\frac{8}{12}} = \frac{2}{3} \underset{\div 4}{}$

| Simplest form | A fraction in which the numerator and denominator cannot both be divided by any number except 1.

You simplify fractions by dividing the numerator and denominator by the same number. | $\frac{3}{4}$ is in simplest form.

$\frac{9}{12}$ is not in simplest form because the 9 and the 12 can both be divided by 3 to give you $\frac{3}{4}$. 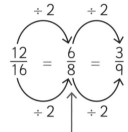

This is not simplest form so divide again. |

1 Look at the diagrams.

   a What fraction of each diagram is shaded yellow?

   b What fraction of each diagram is shaded green?

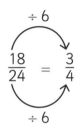

A   B   C

2 Write as many fractions as you can with a denominator of 10 and a value of less than 1.

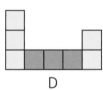

D   E   F

3 Write five fractions with a numerator of 3 and a value of one half or smaller.

4 How many different fractions can you make with the digits 1, 3, 5 and 9? List them.

5 Write each fraction in its simplest form.

   a $\frac{2}{10}$ b $\frac{3}{12}$ c $\frac{6}{8}$ d $\frac{8}{10}$

   e $\frac{9}{18}$ f $\frac{9}{27}$ g $\frac{7}{21}$ h $\frac{4}{12}$

6 Write the following fractions as:

   a quarters: $\frac{2}{8}, \frac{6}{8}, \frac{9}{12}$

   b tenths: $\frac{3}{30}, \frac{20}{50}, \frac{42}{60}$

7 Copy each set of fractions. Write three more equivalent fractions for each set.

   a $\frac{1}{2} = \frac{2}{4} = \frac{3}{6}$ b $\frac{2}{5} = \frac{4}{10} = \frac{6}{15}$ c $\frac{3}{8} = \frac{6}{16} = \frac{9}{24}$

Looking Back

Look at the circle.

Write the fraction shown by each green sector.

Write an equivalent fraction for each one.

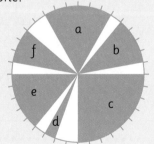

Unit 2 Compare and Order Fractions

Let's Think …

Kendrick wants to arrange the fractions in order from smallest to greatest.
He has this number line divided into eighths.

$$\frac{1}{4} \quad \frac{5}{8} \quad \frac{3}{4} \quad \frac{1}{2} \quad \frac{1}{8}$$

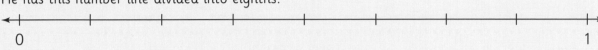

0   1

How can he use this number line to order the fractions?
How else could he do this?

You use the signs $<$, $=$ or $>$ to compare two fractions.

To compare fractions with the same denominators, *look at the numerators*.

$$\frac{3}{8} < \frac{5}{8} \qquad \frac{4}{9} > \frac{1}{9} \qquad \frac{12}{40} < \frac{30}{40}$$

To compare fractions with different denominators, *use equivalent fractions*.

a Compare $\frac{2}{3}$ and $\frac{2}{6}$

Change $\frac{2}{3}$ to get sixths

$$\frac{2}{3} \times \frac{2}{2} = \frac{4}{6}$$

$$\frac{4}{6} > \frac{2}{6}, \text{ so } \frac{2}{3} > \frac{2}{6}$$

b Compare $\frac{2}{3}$ and $\frac{3}{5}$

Change both fractions to get fifteenths

$$\frac{2}{3} \times \frac{5}{5} = \frac{10}{15} \qquad \frac{3}{5} \times \frac{3}{3} = \frac{9}{15}$$

$$\frac{10}{15} > \frac{9}{15}, \text{ so } \frac{2}{3} > \frac{3}{5}$$

You can write sets of fractions in ascending or descending order.

To order sets of fractions, make equivalent fractions with the same denominator.

Arrange these fractions in ascending order: $\frac{5}{6}$, $\frac{1}{2}$, $\frac{1}{4}$, $\frac{2}{3}$

Look at the denominators: 6, 2, 4 and 3
All of these denominators are factors of 12.
Change the fractions to make equivalent twelfths.

$$\frac{5}{6} \times \frac{2}{2} = \frac{10}{12} \qquad \frac{1}{2} \times \frac{6}{6} = \frac{6}{12} \qquad \frac{1}{4} \times \frac{3}{3} = \frac{3}{12} \qquad \frac{2}{3} \times \frac{4}{4} = \frac{8}{12}$$

You can see the order is $\frac{3}{12}$, $\frac{6}{12}$, $\frac{8}{12}$, $\frac{10}{12}$.

Now you can write the original fractions in ascending order.

$$\frac{1}{4}, \frac{1}{2}, \frac{2}{3}, \frac{5}{6}$$

These methods work for mixed numbers as well, but remember to compare the whole numbers first. It does not matter what the fraction part is if the whole numbers are different.

Compare $2\frac{1}{2}$ and $3\frac{3}{4}$

Look at the whole numbers: $3 > 2$, so $3\frac{3}{4} > 2\frac{1}{2}$

1 Say whether the following statements are TRUE or FALSE.

a $\frac{1}{2} > \frac{3}{5}$

b $\frac{6}{8} > \frac{4}{5}$

c $\frac{5}{6} = \frac{7}{8}$

d $\frac{1}{2} < \frac{4}{8}$

e $\frac{2}{3} < \frac{3}{9}$

f $\frac{9}{10} > \frac{2}{3}$

g $\frac{2}{5} < \frac{3}{8}$

h $\frac{2}{3} < \frac{3}{4}$

2 Which of these fractions are greater than $\frac{1}{2}$?

a $\frac{1}{3}$

b $\frac{9}{10}$

c $\frac{7}{13}$

d $\frac{3}{4}$

e $\frac{5}{8}$

3 Which of these fractions are equivalent to $\frac{1}{2}$?

a $\frac{2}{3}$

b $\frac{2}{4}$

c $\frac{3}{6}$

d $\frac{6}{7}$

e $\frac{15}{30}$

4 Compare these fractions using < or > signs.

a $\frac{2}{3}$ and $\frac{3}{4}$

b $\frac{6}{10}$ and $\frac{4}{5}$

c $\frac{1}{2}$ and $\frac{1}{3}$

d $\frac{1}{3}$ and $\frac{5}{12}$

e $\frac{2}{3}$ and $\frac{5}{6}$

f $\frac{4}{5}$ and $\frac{17}{20}$

g $\frac{2}{5}$ and $\frac{41}{100}$

h $\frac{1}{4}$ and $\frac{2}{5}$

5 Arrange each set of fractions in ascending order.

a $\frac{1}{2}, \frac{3}{8}, \frac{1}{3}, \frac{1}{6}, \frac{1}{4}$

b $\frac{1}{2}, \frac{9}{10}, \frac{4}{5}, \frac{1}{10}$

c $\frac{3}{4}, \frac{7}{12}, \frac{1}{3}, \frac{5}{6}$

6 Arrange each set of fractions in descending order.

a $\frac{2}{3}, \frac{1}{2}, \frac{5}{6}, \frac{7}{12}$

b $1\frac{3}{4}, 1\frac{4}{5}, 1\frac{6}{10}, 1\frac{1}{2}$

c $1\frac{3}{4}, 1\frac{7}{10}, 1\frac{2}{5}, 1\frac{4}{5}$

7 $\frac{3}{4}$ and $\frac{6}{8}$ are equivalent fractions.

a How could you show someone that even though the fractions are equivalent, they may not represent the same amount?

b How many fractions equivalent to $\frac{3}{4}$ is it possible to make? Explain your answer.

Looking Back

1 Compare these fractions using <, = or >.

a $\frac{2}{8}$ and $\frac{1}{4}$

b $\frac{1}{5}$ and $\frac{2}{10}$

c $\frac{3}{4}$ and $\frac{7}{8}$

d $\frac{3}{4}$ and 1

2 Write each set of fractions in descending order.

a $\frac{1}{2}, \frac{2}{3}, \frac{3}{4}, \frac{2}{6}$

b $1\frac{1}{8}, 1\frac{1}{5}, 1\frac{1}{12}, 1\frac{1}{2}$

Topic Review

What Did You Learn?

- You write fractions using a numerator and denominator like this: $\frac{2}{3}$.
- The denominator shows how many parts the whole is divided into.
- The numerator tells you how many parts of the whole you are working with.
- Equivalent fractions have the same value.
- You can find equivalent fractions if you multiply or divide the numerator and denominator by the same number.
- You simplify fractions by dividing the numerator and denominator by the same number.
- To order and compare fractions you can use equivalent fractions.

Talking Mathematics

Can you solve these riddles?

- I am a proper fraction. I am equivalent to $\frac{3}{5}$ but my denominator is 20. What am I?
- I am a fraction equivalent to $\frac{3}{4}$. My denominator is greater than 16 but less than 21. What am I?
- I am a mixed number between 2 and 3. The numerator and denominator of my fraction add up to 21. In simplest form it would be $\frac{3}{4}$. What am I?

Quick Check

1 What is the simplest form of each fraction?

  a $\frac{6}{10}$ b $\frac{16}{24}$ c $\frac{20}{25}$ d $\frac{15}{27}$ e $\frac{35}{100}$

2 One fraction in each set is not equivalent to the others. Which one is it?

  a $\frac{4}{18}, \frac{8}{36}, \frac{14}{63}, \frac{2}{9}, \frac{32}{108}$ b $\frac{3}{4}, \frac{10}{12}, \frac{5}{6}, \frac{15}{18}, \frac{45}{54}$

3 Write a mixed number to answer these questions.

  a How many hours is 105 minutes?

  b How many metres is 125 centimetres?

  c How many years is 18 months?

4 Compare the fractions using the $<$, $=$ or $>$ signs.

  a $\frac{5}{10}$ and $\frac{1}{2}$ b $\frac{1}{3}$ and $\frac{4}{10}$ c $\frac{1}{3}$ and $\frac{3}{10}$ d $\frac{4}{10}$ and $\frac{3}{12}$

5 Write each set of fractions in ascending order.

  a $\frac{7}{10}, \frac{3}{12}, \frac{1}{3}, \frac{1}{2}$ b $\frac{2}{8}, \frac{1}{3}, \frac{1}{2}, \frac{9}{12}$ c $1\frac{2}{5}, 1\frac{3}{10}, 1\frac{6}{10}$ d $2\frac{1}{2}, 1\frac{1}{4}, 3\frac{5}{12}, 1\frac{1}{2}$

Topic 6 Mass Workbook pages 16–18

▲ We use scales to measure mass. Which of these kinds of scales have you seen or used? What kinds of everyday items would you weigh using each of these? How do you think they work?

When someone asks, 'How much does it **weigh**?' they are talking about **mass**. Mass is the amount of matter in an object. We also think of mass as how heavy something is. In this topic, you are going to use standard units to estimate and measure mass.

Getting Started

1 Your teacher will give you a crayon and a book. Work in groups.

a Find two items that have about the same mass as the crayon.

b Find two items that have about the same mass as the book.

c Find something lighter than the crayon.

d Find something heavier than the book.

You can use the table in your Workbook to record your items.

2 At home, find five products in the fridge or grocery cupboard that have their mass marked on the bottle or box. Copy and complete this table with the information you find. You may notice that some products are marked in metric units: grams (g) and kilograms (kg), while others use customary units: pounds (lb) and ounces (oz).

Product	Mass on Label	Metric or Customary?

Unit 1 Standard Units of Measure

Let's Think ...

Which units would you use to measure the mass of these?

- Light items such as a piece of paper or a pen.
- Heavier items such as a table or chair.
- A person.
- A car.

The standard unit of measure for mass is the *gram* (g). A paperclip *weighs* about 1 gram. Some things are even lighter than this. We measure them in milligrams. We measure heavier things in kilograms (kg) and tonnes (t). These units form part of the metric system of measurement.

$$1\,g = 1\,000\,mg \qquad 1\,kg = 1\,000\,g \qquad 1\,t = 1\,000\,kg$$

Customary units of measurement are an older system of units. When you see measurements in ounces (oz) and pounds (lb), these are customary measures. The customary system also has a unit called a ton. It is not the same as a metric tonne!

$$1 \text{ pound } (1\,lb) = 16 \text{ ounces } (oz) \qquad 1 \text{ ton } = 2\,000\,lb$$

To convert from a bigger unit to a smaller unit, use multiplication.

To convert from a smaller unit to a bigger unit, use division.

Work out these conversions.

$5\,kg = \boxed{}\,g$

There are $1\,000\,g$ in $1\,kg$.

$5\,kg = 5 \times 1\,000\,g$

$ = 5\,000\,g$

So $5\,g = 5\,000\,g$

$2\,500\,kg = \boxed{}\,t$

There are $1\,000\,kg$ in $1\,t$.

$2\,500 \div 1\,000 = 2.5$

So $2\,500\,kg = 2.5\,t$

1 Choose the most reasonable mass for each object.

a

a lettuce leaf
2 g or 2 kg?

b

a pickup truck
100 kg or 1 t?

c

a whole mango
120 g or 120 mg?

d

schoolbag full of books
30 g or 3 kg?

e

one tablet
500 mg or 500 g?

f

a bag of oranges
$\frac{1}{2}$ t or 5 kg?

g

a block of butter
1 lb or 1 kg?

h

a slice of bread
1 kg or 1 oz?

i

bunch of bananas
1 lb or 1 oz?

2 What pattern do you notice in the relationships between the metric units?

3 Work these out.

   a 3 g = _____ mg   b 2 kg = _____ g   c 4 t = _____ kg

4 How did you work out your answers to question 3?

5 Now work these out.

   a 2 000 mg = _____ g   b 5 000 g = _____ kg   c 3 000 kg = _____ t

6 How did you work out your answers to question 5?

7 Express the following in g.

   a 3 kg   b $2\frac{1}{2}$ kg   c 5.25 kg   d 6.1 kg

8 Express the following in kg.

   a 450 g   b 7 520 g   c 19 000 g   d 11 439 g

9 Work in pairs. Discuss how to work out the following.

   a 2 lb = _____ oz   b 5 lb = _____ oz
   c 20 oz = _____ lb _____ oz   d 48 oz = _____ lb _____ oz

10 Why do you think it is easier to convert metric units than customary units?

11 Solve these word problems.

   a James needs to move books from his classroom to the library. The total mass of the books is
      45 kg. He has a crate that can carry at most 8 kg at a time. How many trips must he make?

   b A farm produces 3.5 metric tons of mangoes. They sell the mangoes in 5 kg boxes to the
      supermarkets. How many boxes can they make?

   c A packet of flour is 2.5 kg. The baker uses 250 g for each batch of muffins. How many
      batches can she make from one packet?

Looking Back

1 How many grams in a kg?
2 How many kg in a tonne?
3 550 g = ☐ kg. Do you use division or
   multiplication to work it out?

4 8.5 kg = ☐ g. Do you use division or
   multiplication to work it out?
5 Create two of your own word problems
   involving g or kg. Work them out and explain
   how you reached the answer.

Topic Review

What Did You Learn?

- Mass is the amount of matter in an object.
- You measure mass using standard units.
- The main metric units for measuring mass are grams (g), kilograms (kg) and tonnes (t).
- Customary measures include ounces (oz), pounds (lbs) and tons.
- You use multiplication to change kg to g.
- Use division to change g to kg.

Talking Mathematics

1 The 'kilo' part of kilogram means 'thousand'. Explain how the kilogram got its name.

2 The word 'pound' has some other meanings. Use a dictionary to look it up and find at least two other meanings.

3 Which is the odd word out in each set? Give a reason for your answer.

a	mass	heaviness	weigh	length
b	grams	tons	tonnes	kilograms

Quick Check

1 How many kilograms in a tonne?

2 How many lbs in a ton?

3 Name three metric units used for measuring mass.

4 Name three customary units used for measuring mass.

5 Which of the following might have a mass of 72 kg: a baby; a car; or a man?

6 Which of the following might have a mass of 120 g: a textbook; an apple; or a pen?

7 Name something you might find at home that could have a mass of 1 lb.

8 Write the following in grams.

  a $\frac{1}{2}$ kg   b 2 kg   c 5.25 kg

9 Write the following in kilograms.

  a 1000 g   b 8 500 g   c 4 295 g

10 Leroy decides to go on a weight loss plan. His starting mass is 123 kg. He aims to lose 500 g per week. His target mass is 90 kg. How long will it take him to reach his target mass?

Topic 7 Decimals Workbook pages 19–21

▲ This is a stick insect. This one is about 12.5 cm long, but the longest one ever found in the world is 62.4 cm or 0.624 metres long! These measurements are all decimals. How do you know that a number is a decimal? Tell your partner.

Key Words
decimal
place value
decimal point
digit
tenths
hundredths
thousandths
equivalent
convert

You worked with **decimals** and decimal **place value** last year. You know that when you write a decimal, you use the **decimal point** to separate the whole number part of a number from the decimal fraction part. The position of the digits after the decimal point allows you to work out their value.

In the decimal 1.345, the 1 is a whole number, the 3 is 3 tenths or $\dfrac{3}{10}$, the 4 is 4 hundredths or $\dfrac{4}{100}$ and the 5 is 5 thousandths or $\dfrac{5}{1\,000}$.

You use the word 'and' in place of the decimal point when you say decimals or write them in words. So, for 1.345 you would say 'one and three hundred forty-five thousandths'.

Getting Started

1 A packet of sugar has a mass of 2.5 kg on the label.
  a What does the 2 stand for?   b What does the 5 mean?

2 Jerome weighs an eraser and finds it has a mass of 20.43 grams.
  a Is this more or less than 20 grams?
  b What is the value of the digit 3 in the measurement? How do you know that?

3 The price of a T-shirt is $9.99.
  a Why is this money amount written as a decimal?
  b How would you write 85 cents as a decimal?

4 How much money would you have if you had $\dfrac{1}{10}$ of a dollar, 3 whole dollars and $\dfrac{7}{100}$ of a dollar? Write the total as a decimal.

Unit 1 Revisit Decimal Place Value

Let's Think ...

Shaundra has chosen one of the numbers in the box.

| 0.632 | 0.636 | 0.654 | 2.341 | 1.537 | 0.63 | 0.535 | 0.637 |

Read the clues. Which number did she choose?

- It is less than 1.
- It has 3 in the hundredths place.
- It is more than 0.631.
- It has more than 5 thousandths
- There is an odd number in the thousandths place.

Do you remember how to read and write decimals to thousandths?
Read through this information to refresh your memory.
You extend the place value table to the right to show decimals.

Ones	.	Tenths $\left(\frac{1}{10}\right)$	Hundredths $\left(\frac{1}{100}\right)$	Thousandths $\left(\frac{1}{1000}\right)$
0	.	4	8	2
2	.	0	7	8

The decimal point separates the whole numbers from the fractional parts.
The position of the digits tells you their value.

Look at these two decimals.

This decimal is less than 1   This decimal is greater than 1

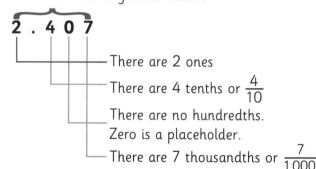

0 . 2 4 6   2 . 4 0 7

There are no ones.
Zero is a placeholder.

There are 2 tenths or $\frac{2}{10}$

There are 4 hundredths or $\frac{4}{100}$

There are 6 thousandths or $\frac{6}{1000}$

There are 2 ones

There are 4 tenths or $\frac{4}{10}$

There are no hundredths.
Zero is a placeholder.

There are 7 thousandths or $\frac{7}{1000}$

You can write these decimals in expanded form like this:

$$0 + \frac{2}{10} + \frac{4}{100} + \frac{6}{1000}$$

$$2 + \frac{4}{10} + \frac{0}{100} + \frac{7}{1000}$$

In words, you write:

Two hundred forty-six thousand**ths**

Two **and** four hundred seven thousand**ths**.

You say '*and*' in place of the decimal point.

1 Say each decimal in words then write the value of the blue digit in each number.

  a 0.345   b 0.808   c 3.041   d 2.409   e 3.098

2 Write these fractions as decimals.

  a $\frac{3}{10}$ b $\frac{42}{100}$ c $\frac{9}{1\,000}$ d $2\frac{3}{10}$ e $9\frac{12}{1\,000}$

3 Write these numbers as decimals.

  a two hundred seven thousandths   b two and seven hundredths

  c nine and four hundred thirty-one thousandths   d three and nine tenths

4 Write the following as decimals.

  a $3 + 0.3 + 0.07 + 0.001$ b $10 + 5 + 0.8 + 0.005$

  c $0.009 + 0.6 + 2$ d $2 + \frac{5}{100} + \frac{3}{10} + \frac{9}{1\,000}$

5 Look at these cards.

Using all five of the cards each time, make:

  a a number less than 1   b a number greater than 20

  c the closest number to 27 that is possible   d the greatest possible number

  e the smallest possible number.

6 Marvin has 1 kilogram of sugar. He uses 180 grams in a recipe.

  a Write 180 grams as a decimal fraction of the whole amount of sugar.

  b After using the 180 grams, Marvin puts 0.75 kg of the sugar into a tin. Write 180 grams in kilograms as a decimal fraction.

Looking Back

1 Is one thousandth bigger or smaller than one hundredth? Explain your answer.

2 How many thousandths make one whole?

3 How many thousandths make one hundredth?

4 Why can you not write five thousandths as 0.5?

Unit 2 Compare and Order Decimals

Let's Think ...

Fairyflies are small wasps. The average length of a fairyfly is 0.139 mm.

A species of beetle found in Colombia is 0.32 mm long.

Which is longer, the fairyfly or the beetle?

Is a fairyfly longer or shorter than $\frac{1}{10}$ of a millimetre?

How did you decide?

You use place value to compare decimals.

The flow diagram shows the steps you follow to do this.

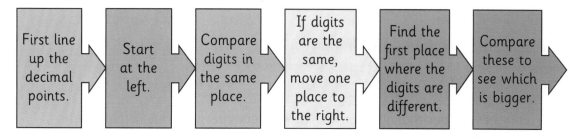

First line up the decimal points. ➤ Start at the left. ➤ Compare digits in the same place. ➤ If digits are the same, move one place to the right. ➤ Find the first place where the digits are different. ➤ Compare these to see which is bigger.

Read through the examples to make sure you understand.

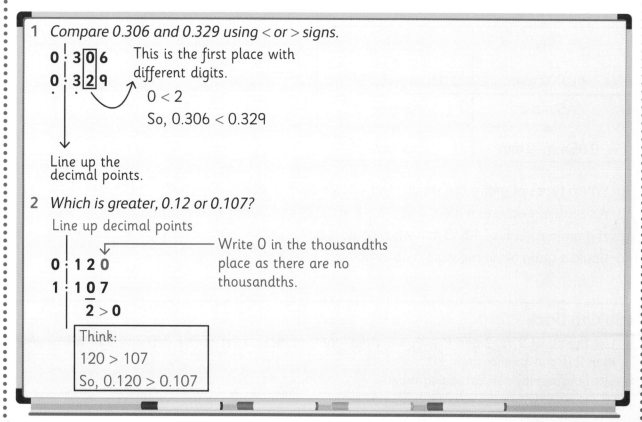

1 Compare 0.306 and 0.329 using < or > signs.

0 . 3 [0] 6

0 . 3 [2] 9

This is the first place with different digits.

0 < 2

So, 0.306 < 0.329

Line up the decimal points.

2 Which is greater, 0.12 or 0.107?

Line up decimal points

0 . 1 2 0

1 . 1 0 7

Write 0 in the thousandths place as there are no thousandths.

$\underline{2} > 0$

Think:

120 > 107

So, 0.120 > 0.107

1 Write each pair of numbers with <, = or > between them.

   a 0.5 and 0.05   b 1.25 and 1.255   c $\frac{7}{10}$ and 0.07

   d 1.500 and 1.5   e 1.03 and 1.303   f 0.12 and 0.125

   g 7 and 7.02   h 0.87 and 0.870   i 0.002 and 0.02

2 Write the smaller number in each pair.

   a 4.32 and 4.23   b 21.34 and 21.09   c 12.234 and 12.243

   d 9.013 and 9.12   e 4.2 and 4.201   f 9.05 and 9.1

3 Arrange these numbers in descending order.

   a 4.44,   4.49,   4.09,   4.99,   4.105

   b 13.21,   11.22,   13.03,   13.2,   11.231

4 Arrange these lengths in ascending order.

   a 1.02 m   1.002 m   2.001 m   2.010 m   1.111 m

   b 7.32 m   7.325 m   7.3 m   7.53 m   7.35 m

5 Which of these measurements are greater than 0.125 cm?

   0.5 cm   0.3 cm   0.13 cm   0.12 cm

6 The mass of an insect is a decimal number with three places between 4.508 grams and 4.515 grams. What could it possibly be?

7 Soil grains are named by size. The sizes of clay, silt, sand and gravel grains are given here.

Clay	Silt
•	•
<0.002 mm	0.002 mm–0.05 mm
Sand	Gravel
•	•
0.05 mm–2 mm	>2 mm

   a Which types of grains are smaller than 5 hundredths of a millimetre?
   b A scientist measures a grain and finds it is 0.035 mm. What type of grain is it?
   c If a grain measures 1.543 mm what could it be?
   d Could a grain of silt measure 0.45 mm?

Looking Back

a Use the digits 2, 5, 3 and 9 with a decimal point to make a set of five decimals that are greater than 2.0 and smaller than 3.0.
b List your numbers in ascending order.

Unit 3 Convert Between Fractions and Decimals

Let's Think …

Work with a partner.

If $\frac{1}{10} = 0.1$ and $\frac{2}{10} = 0.2$, what is the decimal for each of these?

$\frac{3}{10}$ $\frac{4}{10}$ $\frac{5}{10}$ $\frac{8}{10}$ $\frac{9}{10}$ $\frac{10}{10}$?

What is the rule for writing tenths as decimals?

How would you write 0.5 as a fraction in simplest terms?

In mathematics you often have to *convert* decimals to fractions or fractions to decimals.

Converting decimals to fractions is simple if you think of them as *tenths, hundredths* or *thousandths*. You should always simplify to give the fraction in its simplest terms.

Write 0.8 as a fraction.

0.8 is 8 tenths or $\frac{8}{10}$ $\frac{8}{10} \div \frac{2}{2} = \frac{4}{5}$

So, $0.8 = \frac{4}{5}$

Convert 0.25 to a fraction in simplest terms.

0.25 is 25 hundredths or $\frac{25}{100}$

$\frac{25}{100} \div \frac{5}{5} = \frac{5}{20}$ $\frac{5}{20} \div \frac{5}{5} = \frac{1}{4}$

So, $0.25 = \frac{1}{4}$

A whole number part means your answer will be a mixed number.

Write 1.125 as a fraction.

1.125 is 1 and 125 thousandths or $\frac{25}{1000}$

$\frac{125}{1000} \div \frac{5}{5} = \frac{25}{200}$

$\frac{25}{200} \div \frac{5}{5} = \frac{5}{40}$

$\frac{5}{40} \div \frac{5}{5} = \frac{1}{8}$

Remember to include the whole number part. The answer is $1\frac{1}{8}$

There are two ways to convert fractions to decimals.

Method 1: Write an *equivalent* fraction with a denominator of 10, 100 or 1 000. Then, write the fraction as a decimal.

$\frac{1}{2} \times \frac{5}{5} = \frac{5}{10} = 0.5$ $\frac{2}{5} \times \frac{2}{2} = \frac{4}{10} = 0.4$ $\frac{1}{4} \times \frac{25}{25} = \frac{25}{100} = 0.25$

Method 2: Divide the numerator by the denominator.

Write 0 after the decimal place to make the division easier.

Line up the decimal points in your calculation.

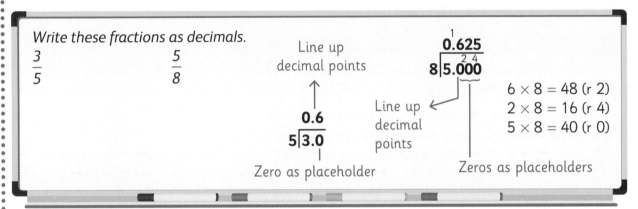

1 Write these decimals as fractions or mixed numbers.
   a 0.004   b 0.06   c 8.003   d 0.876   e 5.005

2 Write these fractions or mixed numbers as decimals.
   a $\dfrac{12}{100}$ b $\dfrac{12}{1\,000}$ c $1\dfrac{2}{100}$ d $\dfrac{12}{1\,000}$ e $\dfrac{500}{1\,000}$ f $3\dfrac{90}{1\,000}$

3 Convert to decimals. Check your answers using a calculator.
   a $\dfrac{9}{10}$ b $\dfrac{1}{5}$ c $3\dfrac{1}{2}$ d $2\dfrac{3}{5}$ e $\dfrac{5}{5}$ f $2\dfrac{4}{5}$ g $10\dfrac{4}{5}$ h $\dfrac{2}{5}$

4 Convert each set of fractions to decimals and then write the fractions in ascending order.
   a $\dfrac{2}{20}$, $\dfrac{2}{5}$, $\dfrac{7}{10}$, $\dfrac{1}{4}$
   b $\dfrac{80}{200}$, $\dfrac{3}{4}$, $\dfrac{1}{2}$, $\dfrac{1}{5}$
   c $\dfrac{4}{5}$, $\dfrac{3}{10}$, $\dfrac{9}{20}$, $\dfrac{9}{100}$

5 In a spelling contest, students are given a set of words to spell and they get a score based on how many they get correct. Students in different grades get different numbers of words.

   These are the results for three students in School A:

   Jaynae $\dfrac{29}{40}$ Michael $\dfrac{29}{50}$ Sheldon $\dfrac{17}{25}$

   a Convert the scores to decimals and decide who would get first, second and third place.
   b Decide who would get first, second and third place in School B.
      School B: Nisha $\dfrac{50}{80}$ Elayne $\dfrac{30}{50}$ Peter $\dfrac{128}{200}$
   c Which student would come first overall? Why?

Looking Back

How can you use decimals to work out whether $\dfrac{80}{25}$ is greater or less than $\dfrac{31}{40}$?

Topic Review

What Did You Learn?

- A decimal fraction is written with a decimal point separating the whole number part from the fractional part.
- Numbers to the left of the decimal point are whole numbers.
- Numbers to the right of the decimal point are fractions of a whole.
- The place value table is extended to include decimals using tenths, hundredths and thousands.
- $\frac{1}{10} = 0.1$, $\frac{1}{100} = 0.01$ and $\frac{1}{1\,000} = 0.001$.
- You use place value to compare the size of decimals. Compare digits from left to right.
- You can convert decimals to fractions by writing them with a denominator of 10, 100 or 1 000 and then simplifying them.
- You can convert fractions to decimals by making equivalent fractions with a denominator of 10, 100 or 1 000 and then writing them as decimals. You can also divide the numerator by the denominator.

Talking Mathematics

- Joshua says $0.7 = 0.70 = 0.700$. Explain why he is correct.
- Micah says $\frac{67}{1\,000}$ is 0.67. Explain why he is incorrect.
- Tonique asks you how you can use decimal fractions to compare ordinary fractions with different denominators. What would you tell her?

Quick Check

1 Write these fractions as decimals.

   a $\frac{4}{10}$ b $\frac{87}{1\,000}$ c $\frac{9}{100}$ d $\frac{120}{1\,000}$ e $\frac{2}{1\,000}$

2 Write each quantity as a decimal fraction.

   a $1\frac{89}{100}$ b $3\frac{2}{5}$ c $\frac{43}{1\,000}$ d $3\frac{1}{4}$ e \$10 and 3 cents   f $3\frac{12}{40}$ kilometres

3 Compare 0.34 and 0.309 using $<$ or $>$.

4 The world's smallest insect is a fairyfly which is 0.139 mm long.

   a What is the value of the 1 and the 9 in this number?

   b Another fairyfly is $\frac{1}{100}$ of a millimetre longer. How long is that one?

   c A fairyfly 0.143 mm long is 100 times shorter than an ant. How long is the ant?

Topic 8 Classifying Shapes

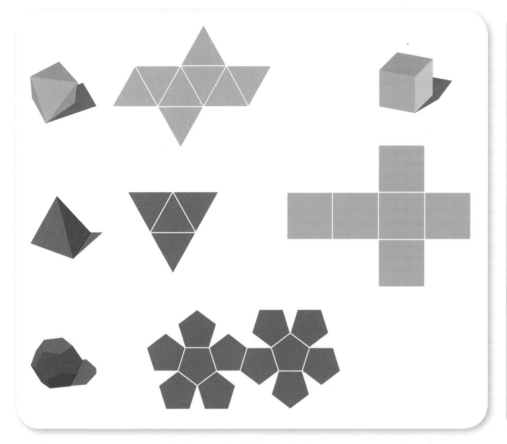

Key Words

regular polygon
irregular polygon
quadrilateral
square
parallelogram
rhombus
kite
trapezium
rectangle
sphere
cone
cylinder
cube
triangular pyramid
square pyramid
rectangular prism
triangular prism

▲ Can you name any of these solid shapes? Can you name the shapes
  that make up their faces?

As you learn more about shapes in two and three dimensions, you will
start to realize the importance of geometric shapes in our lives. Do
you think we can construct a building without a plan using geometric
properties and relationships? In this topic, you will learn the difference
between an open and a closed curve, and when a curve is simple. You will
also revisit and learn more about **quadrilaterals**. Finally, you will learn to
identify and name certain solid shapes.

Getting Started

1 Look around you and try to find ten examples of polygons and solid shapes that you have
   learned about in the past.
2 Now try to imagine your classroom without any of these shapes.
   What would it look like?
   How would the space you are in be different?

Unit 1 Polygons

Let's Think ...

- Can you remember what a polygon is?
- Can you remember the names of any polygons you have learned about?

- What do you notice about the three shapes on the left and the three shapes on the right?

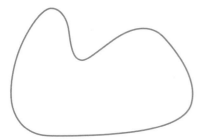
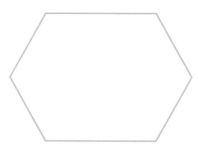

*A **simple closed curve** is a curve that joins up so there are no endpoints and does not intersect itself, for example a circle or an oval. If you draw the shape, you will end up where you started. A simple closed curve is a plane shape.*

*A **simple open curve** is a curve with endpoints (the ends do not join up) which does not intersect itself.*

A curve is not simple if it intersects itself, like the open and closed curves shown below.

A *polygon* is a closed plane figure with at least 3 straight sides. Polygons are named according to the number of sides and angles they have. The table below shows you the names as well as examples of polygons with up to ten sides.

Name	Sides	Angles	Example
Triangle	3	3	
Quadrilateral	4	4	
Pentagon	5	5	
Hexagon	6	6	
Heptagon	7	7	
Octagon	8	8	
Nonagon	9	9	
Decagon	10	10	

A *regular* polygon has all angles equal and all sides equal, otherwise it is *irregular*.

Look at the following regular and irregular triangles, quadrilaterals and pentagons.

Regular	Irregular

1 Indicate whether the following numbers and letters form curves that are open or closed, simple or not simple.

J	8	O	4	X
a	b	c	d	e

2 Use string or bits of wool and glue to make a poster about open and closed curves. Make some curves simple and others not simple.

3 Explain why the following shapes are not polygons.

 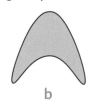 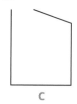

a	b	c	d

4 State whether the following polygons are regular or irregular.

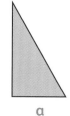 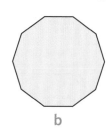 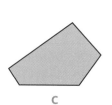  

a	b	c	d	e

5 Classify the following polygons as regular or irregular, followed by their names.

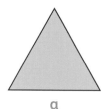   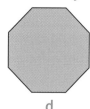 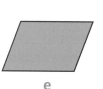

a	b	c	d	e

Looking Back

Make a drawing of each of the following:
- a closed curve that is not simple
- an open curve that is simple
- a regular polygon
- an irregular polygon.

Unit 2 Quadrilaterals

Let's Think …

Did you recognize any of the quadrilaterals in the previous unit?
Are there other quadrilaterals apart from the square that have a consistent shape or regular features?

A quadrilateral is a polygon with four sides and four angles.

A square is the only regular quadrilateral.

It has four equal sides and four equal angles. Each angle is 90°.

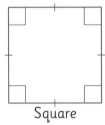

Square

Remember that the small square marking on an angle means that it is a right angle. The small lines on the sides of the square show that the sides are equal in length.

There are other special quadrilaterals that you need to know.

A parallelogram is a quadrilateral with opposite sides parallel and equal in length. Also, the opposite angles are equal. *Pay attention to how the two equal sides are marked using little lines.* *The equal angles are marked with the same lower-case letter.*	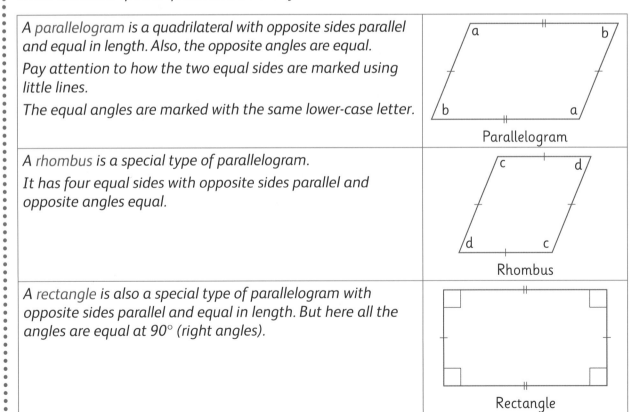 Parallelogram
A rhombus is a special type of parallelogram. *It has four equal sides with opposite sides parallel and opposite angles equal.*	Rhombus
A rectangle is also a special type of parallelogram with opposite sides parallel and equal in length. But here all the angles are equal at 90° (right angles).	Rectangle

A *kite* is a quadrilateral that has two sets of sides next to each other that are equal, as you can see in the diagram below. The two angles where the pairs of sides meet are equal as well.

Kite

A *trapezium* has one pair of opposite sides parallel. In the USA, this shape is called a trapezoid.

Trapezium

1 How many right angles does a square have?

2 Which one is not a name for the figure below?

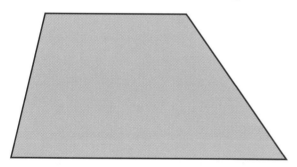

- polygon
- parallelogram
- quadrilateral
- trapezoid

3 Look at these four statements.

A   It has four equal sides.

B   It is a parallelogram.

C   It is a quadrilateral.

D   It always has four right angles.

Which statements best describe a rhombus?

- Just A and C.
- A, B and C.
- A, B and D.
- A, B, C and D.

4 Winston's teacher asked him to draw a quadrilateral with all sides equal.
Choose all the shapes from the box that could he draw.

Square	Rectangle	Parallelogram
Kite	Rhombus	Trapezium

5 Which of the following is a parallelogram?

A Square

B Rectangle

C Rhombus

D All of the above

6 Look at the kite below.

a What is the size of ∠ABC?

b Tell your partner how you worked out the size of angle ABC.

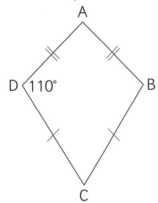

7 What is the difference between a trapezium and a trapezoid?

8 Find four popsicle sticks or plastic drinking straws that are all the same length.
Join them together with sticky putty or split pins to make a quadrilateral.
Experiment to see how many different types of quadrilateral you can model by
moving the sticks or straws to change the angle at the corners.

Looking Back

1 Why can you say that a rectangle is a special type of parallelogram?

2 Why can you say that a square is a special type of rectangle?

3 Can you also say that a square is a type of:

• parallelogram

• rhombus?

4 Draw a diagram to show the relationships between a parallelogram, rectangle, rhombus
and square.

Unit 3 Solids

Let's Think …

Remember that a plane shape is two-dimensional. This means that it has length and width; for example, a rectangle. Now what would happen if you add height to that plane shape? Can you imagine what the rectangle would look like?

Solid objects are three-dimensional because they have length, width and height. A sphere, a cone, a cube and a pyramid are examples of solid objects.

Solid objects have faces, edges and vertices.

- *A **face** is a flat side of a solid.*
- *An **edge** is where two faces of a solid meet.*
- *A **vertex** is a point where three or more faces of a solid meet.*

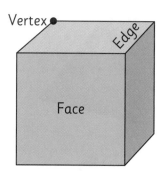

Here is some information about several special solid objects.

A sphere *is a round three-dimensional object. Balls, oranges and marbles are all examples of spheres. Since a sphere has only a curved surface and no flat surface, it has no faces, edges or vertices.*	
A cone *has one flat side, its base, and a curved surface that ends in a point. The point is called an apex. A cone has one face and no edges or vertices. A traffic cone, an ice-cream cone and a party hat are examples of cones.*	

A *cylinder* has two faces, flat circles, and one curved surface. Since the two faces do not meet, a cylinder has no edges and no vertices. Examples of cylindrical objects include canned food tins, batteries and glue sticks.

A *cube* is a solid object with six square faces. All the surfaces are flat, so there are twelve edges and eight vertices. Cubes make very good six-sided dice, as they are regular in shape and each face is the same size. In the diagram of a cube, three of the faces are labelled, all the lines are edges and all the vertices are marked with a dot.

A *pyramid* is a solid object with a polygon base and triangular sides that meet at a single point.

A *triangular pyramid* has a base that is a triangle and three sides that are also triangles. Therefore, it has four faces that are all triangles. It has four vertices and six edges.

The huge pyramids in Egypt are *square pyramids*. This means that the base is a square and the sides are triangles. A square pyramid has five faces, the base and the four sides. It has five vertices and eight edges.

A *prism* is a solid object that has two identical bases and all the sides are parallelograms. The name of the prism depends on the shape of the bases. The prism in the diagram is a *rectangular prism*.

The two ends or bases are rectangles. The four sides are rectangles, which are parallelograms. A rectangular prism has six faces, eight vertices and twelve edges.

This prism is a *triangular prism*. The two bases are triangles and the three sides are parallelograms. That makes five faces, six vertices and nine edges.

1 Write down the name of each solid object.

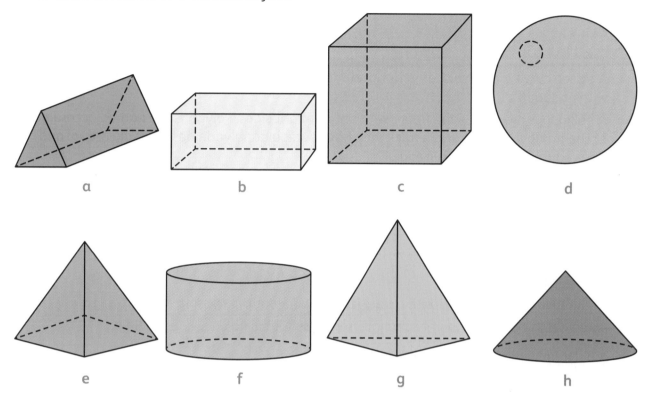

a   b   c   d

e   f   g   h

2 Have another look at the solid objects above, then write down the names of all the solids that apply to each statement below.

a Each edge is perpendicular (at right angles) to four other edges.

b There are no perpendicular edges.

c There are no vertices.

d There are six edges.

e There are three side edges.

f There are opposite faces that are parallel.

g All the side faces are triangles.

h There are no edges.

3 Work in groups. Your teacher will give you nets of solids. Construct the solids and then mark the faces, edges and vertices. As a group, make a chart about your findings.

Looking Back

Why does a sphere not have any edges or vertices?

Why does a cone not have any edges or vertices?

Why does a cylinder not have any edges or vertices?

Why does a rectangular prism have the same number of faces, edges and vertices as a cube?

Topic Review

Talking Mathematics

Read the following sentences with a partner and work out the missing words.

a A curve that has no endpoints and intersects itself is called a ____ ____ curve.

b A polygon with six sides that are equal is called a ____ ____.

c A quadrilateral is a polygon with ____ sides.

d A ____ is a quadrilateral with opposite sides parallel and equal in length, and opposite angles equal.

e A ____ is a quadrilateral that has two sets of sides next to each other that are equal, and the two angles where the pairs of sides meet equal.

f Figures that have length, width and height are ____.

g The flat side of a solid is called a ____. Two faces of a solid meet at an ____. Three or more faces of a solid meet at a point called a ____.

h A ____ is a solid with six square faces.

i A solid with a square base and four triangular sides is called a ____.

j A ____ is a solid with two bases exactly the same and all the sides parallelograms.

Quick Check

1 How many sides and angles does a decagon have?

2 Describe an irregular polygon.

3 Which quadrilaterals are all parallelograms?

4 How would you describe a trapezium?

5 Which solids have no edges and no vertices?

6 Which solid has no faces?

7 How would you describe a pyramid?

Topic 9 Rounding and Estimating Workbook pages 27–28

Key Words
round
nearest
estimate
place
digit
leading figure
approximate

▲ These bananas were on sale at an agricultural show. About how many would you say there are? Tell your partner how you decided.

You already know how to **round** whole numbers and decimals to a given number of **places**. Rounded values are often used in everyday life when it is not important to have an exact value. You may have heard people say things like: 'I live about 2 kilometres from the beach', 'there were about 50 people at my party' or 'I spend around $40.00 a week on transport'.

These are **estimated** values, but they are not wild guesses. If someone says they live about 2 kilometres from the beach, it means that the real distance is closer to 2 km than it is to 1 km or 3 km. Rounding and estimating are very useful in mathematics too. You are going to practice these skills so that you can use them effectively.

Getting Started

1 Nina says 'there are about 400 children in my school'.
   a Does this mean there are exactly 400 children in the school? Explain why or why not.
   b Assuming Nina rounded the number of children correctly, could there be 350 children in the school? Why?
   c Could there be 450 children in the school? Why?

2 Mr Samuels weighs some bananas and finds they have an average mass of 0.142 kilograms.
   a Is he correct if he says the average mass is about 0.1 kilograms? Why?
   b About how many bananas would you expect there to be in 1 kilogram? Why?

Unit 1 Round Whole Numbers and Decimals

Let's Think …

A decimal has three digits: 3, 4 and 6 and two decimal places. When it is rounded to the nearest tenth, it is half of 8.8.

Work out what the decimal could be.

The rules for rounding whole number and decimals are the same.

- *Find the digit in the rounding place.*

- *Look at the digit to the right of this place.*

- *If the digit to the right is 0, 1, 2, 3, or 4, leave the digit in the rounding place as it is. If the digit to the right is 5, 6, 7, 8, or 9, add 1 to the digit in the rounding place.*

- *Change all the digits to the right of the rounding place to 0.*

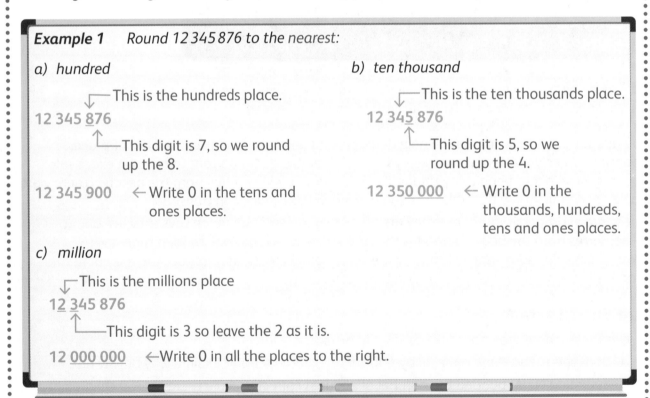

Example 1   Round 12 345 876 to the nearest:

a) hundred

This is the hundreds place.

12 345 8<u>7</u>6

This digit is 7, so we round up the 8.

12 345 900   ← Write 0 in the tens and ones places.

b) ten thousand

This is the ten thousands place.

12 34<u>5</u> 876

This digit is 5, so we round up the 4.

12 350 000   ← Write 0 in the thousands, hundreds, tens and ones places.

c) million

This is the millions place

1<u>2</u> 345 876

This digit is 3 so leave the 2 as it is.

12 000 000   ←Write 0 in all the places to the right.

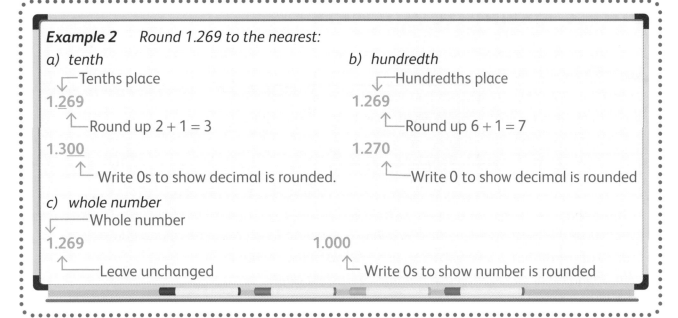

Example 2   Round 1.269 to the nearest:

a) tenth

Tenths place

1.2<u>6</u>9

Round up 2 + 1 = 3

1.<u>3</u><u>00</u>

Write 0s to show decimal is rounded.

b) hundredth

Hundredths place

1.26<u>9</u>

Round up 6 + 1 = 7

1.2<u>7</u>0

Write 0 to show decimal is rounded

c) whole number

Whole number

1.269

Leave unchanged

1.000

Write 0s to show number is rounded

1 How would you round these numbers to make them more user-friendly?

  a There are 12 309 students in our school district.

  b Last year 1 340 876 tourists visited the Caribbean during winter.

  c The population of The Bahamas in 2015 was 387 497.

  d A charity raised $1 346.97 to help disabled children.

  e I live 13.765 km from town.

  f A bakery uses 54.826 kilograms of flour every week.

2 Round each decimal to the nearest hundredth, tenth and whole number. Write the three rounded figures next to each other.

  a 16.325   b 75.654   c 10.011   d 172.299   e 45.298

3 Round each amount to the nearest ten cents and to the nearest dollar.

  a $14.72   b $18.09   c $9.56   d $0.97

4 The population of each country below has been rounded to the nearest thousand.

Country	The Bahamas	Barbados	Trinidad	Jamaica	Cuba
Population	388 000	283 000	1 357 000	2 729 000	11 252 000

  a What is the minimum and maximum number of people that could be in each country?

  b Which country has the closest to 3 million people?

Looking Back

Use the figures from the table above.

Round the populations of Trinidad, Jamaica and Cuba to the nearest million.

Which country has a population of 300 000 to the nearest hundred thousand?

Unit 2 Estimate Answers

Let's Think ...

Jayden has $25.00. He wants to buy three items costing $5.89, $3.90 and $15.65.

● How can he estimate to decide whether he has enough money?

● Does he have enough money?

Estimating is a very useful strategy to help you calculate quickly and to help you decide whether your answer is reasonable or not. You can use rounding to estimate an approximate answer.

For most estimated answers you can use leading figure rounding.

This means that you round each number to the first digit.

Example 1

Estimate 468 × 62

500×60 Round each number to the first digit.

$5 \times 6 = 30$ Use the facts you already know.

so $500 \times 60 = 30\,000$

$468 \times 62 \approx 30\,000$ $\approx$ means 'approximately equal to'

When you give an estimated answer, you use the $\approx$ symbol.

Example 2

Estimate 416 + 338 + 147 + 407

$$
\begin{array}{r}
400 \\
300 \\
100 \\
+\ 400 \\
\hline
1\,200
\end{array}
$$
    Round each number to the first digit.

    Add in columns as normal.

$416 + 338 + 147 + 407 \approx 1\,200$

1 Estimate each of the following.

   a $39 + 42$ b $29 + 187$ c $234 + 123 + 97$ d $32\,876 + 13\,087$

   e $499 - 67$ f $148 - 9 + 24$ g 32×12 h 58×22

   i $32\,876 - 12\,909$ j $3.63 + 9.8 + 6.21$ k $18.23 - 10.15$ l $230.44 - 64.53$

2 Mrs Solomon drove 23.47 km on Monday, 38.43 km on Tuesday, 19.56 km on Thursday and 63.2 km on Friday. Estimate how far she drove during the week.

3 Estimate the total cost of the following items in whole dollars.
  a $28.80 + $49.99
  b $11.82 + $12.20 + $23.89 + $55.12
  c $19.99 + $19.99 + $9.90

4 The numbers of cruise passengers passing through a terminal during a week are given in the table.

Day	Monday	Tuesday	Wednesday	Thursday	Friday
Number of passengers	13 689	25 908	18 579	18 798	21 809

  a Use leading figure rounding to estimate the total number of passengers over the five days.
  b Round the passenger numbers to the nearest thousand and then estimate the total number over five days.
  c Compare the two estimates. Which one is a more reliable figure? Why?
  d Use the figures rounded to the nearest thousand and estimate:
    i   how many more passengers passed through on Friday than on Thursday
    ii the number of passengers there were on the two busiest days
    iii the difference in passenger numbers between the busiest and least busy days.

5 There are 86 400 seconds in a day. Estimate how many seconds there are in:
  a two days   b a week.

6 Crowd attendance at a cricket match over a three-day period was 4 146, 5 964 and 7 193.
  a Estimate the total attendance.
  b If tickets cost $9.50 per day, estimate how much money was spent on tickets over the three days.

7 Approximately how many sets of 180 counters could you make from a box containing 24 275 counters?

Looking Back
Use the scale and estimate the distance from:
a Townsville to the Harbour
b Beachville to Cityville by road
c Farmville to Beachville and back, three times a week.

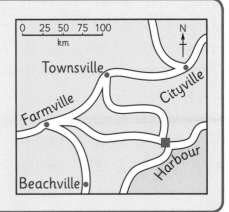

Topic Review

What Did You Learn?

- You can round numbers and decimals to any place using the same rules.
- If the digit to the right of the rounding place is 5 or greater, add one to the digit in the rounding place.
- If the digit to the right of the rounding place is 4 or less, leave the digit in the rounding place unchanged.
- Replace the digits to the right of the rounding place with zeros.
- Rounding is useful for estimating when you do not need an exact value.
- You can round a number to the first digit. This is called the leading figure.
- When you have estimated an answer, you use the ≈ symbol. This means approximately equal to.

Talking Mathematics

1 Give three examples of where you might use or hear rounded numbers in your everyday life.
2 Explain why you get 1.00 when you round 0.99 to the nearest tenth.

Quick Check

1 Round to the nearest thousand.
  a 17 989   b 2 345   c 45 876   d 293 098   e 96 098

2 Round the nearest hundredth.
  a 2.935   b 7.099   c 4.361   d 1.213   e 23.876

3 Round 13 324 987 to the nearest hundred thousand.

4 Round each decimal to the nearest whole number and estimate the answer.
  a 138.22 + 271.66   b 44.501 + 80.89   c 10.57 + 13.2 + 9.99

5 Use rounding to estimate the answers to these calculations.
  a 603 + 715 + 986   b 7 899 − 5 211
  c 24 999 ÷ 2   d 408 × 31

6 Estimate the area of each shape.

9.8 cm

27.9 cm

19.3 cm

21.3 cm

Topic 10 Factors and Multiples Workbook pages 29–32

Key Words

multiple

factor

product

common factor

common multiple

greatest common factor (GCF)

least common multiple (LCM)

composite number

prime number

prime factors

▲ Look at the sheet of stickers. How many stickers are in each row? How many rows are there? How can you use these numbers to find the total number of stickers? How many ways can you find to group the stickers so that there is the same number in each group and none left over?

Last year, you learned that you could multiply two **factors** to find their **product**. This year, you are going to build on what you know about multiplication and division to learn more about factors and **multiples**. You will also learn about the properties of different types of numbers and how you can use these properties to find the factors of any number.

Getting Started

1 What is the product of each of these pairs of factors?

   a 4 and 9   b 2 and 8

   c 1 and 13   d 4 and 4

2 Make a list of all the whole numbers that can divide into 24 without leaving a remainder.

3 Read this information and discuss it with your partner.

These are multiples of 5:	5, 10, 15, 20, 100, 55, 90
These are not multiples of 5:	1, 12, 18, 24, 62, 99

Write in your own words what you think a multiple is.

Unit 1 Factors and Multiples

Let's Think ...

$3 \times 8 = 24$, so 3 and 8 are a factor pair with a product of 24.
How many other factor pairs can you find with a product of 24?
List them.

There are three different pairs of *factors* that result in a *product of 12*.

$1 \times 12 = 12$ $2 \times 6 = 12$ $3 \times 4 = 12$

This shows that a number can have many factors.

1, 2, 3, 4, 6 and 12 are all factors of 12.

A factor of a number is any number that divides into it without a remainder.

2 is a factor of 6 because $6 \div 2 = 3$ (you know that $2 \times 3 = 6$).

4 is not a factor of 6 because $6 \div 4 = 1$ remainder 2. You cannot multiply 4 by a whole number to get a product of 6.

All numbers can be divided by 1, so 1 is a factor of every number.

All numbers can also be divided by themselves (and the result will be 1), so every number is also a factor of itself.

Some numbers share factors.

Factors of 12:   1, 2, 3, 4, 6, 12   Factors of 16:   1, 2, 4, 8, 16

1, 2 and 4 are factors of 12 and factors of 16. We call these *common factors*.

4 is the *greatest common factor (GCF)* of 12 and 16.

GCF is a short way of writing greatest common factor.

To find the GCF, list the factors. Circle the common factors and find the highest number that is common to the sets.

> Find the greatest common factor of 24 and 36.
> Factors of 24:   ①, ②, ③, ④, ⑥, 8, ⑫, 24
> Factors of 36:   ①, ②, ③, ④, ⑥, 9, ⑫, 18, 36
> The GCF of 24 and 36 is 12.

Multiples of a number are found when you multiply that number by any other whole number.

The first ten multiples of 3 are: 3, 6, 9, 12, 15, 18, 21, 24, 27, 30

The first ten multiples of 4 are: 4, 8, 12, 16, 20, 24, 28, 32, 36, 40

12 and 24 are multiples of 3 and multiples of 4. We call these *common multiples*.

The lowest number that is common to both sets is 12.

We say that 12 is the *least common multiple (LCM)* of 3 and 4.

To find the least common multiple of two numbers, list the multiples in order till you get to one that is shared by both.

Find the LCM of 12 and 3.

Multiples of 12: ⓬, 24, 36

Multiples of 3:   3, 6, 9, ⑫ You can stop when you have a common multiple.

The LCM of 12 and 3 is 12.

1 Read each statement. Say whether it is TRUE or FALSE.
  a 5 is a factor of 20.   b 5 is a factor of 24.
  c 9 is a multiple of 3.   d All even numbers are multiples of 2.
  e 10 is a factor of 80.   f 75 is a multiple of 10.

2 Write the first five multiples of each of these.
  a 5   b 9   c 7   d 10   e 2

3 List the multiples of:
  a 4 between 25 and 45   b 100 between 450 and 950.

4 Find the least common multiple of each pair.
  a 6 and 4   b 2 and 9   c 12 and 8   d 2 and 7   e 5 and 8   f 6 and 7

5 Find all the pairs of factors for each number.
  a 4   b 7   c 10   d 14   e 20   f 25   g 28   h 30

6 Find the greatest common factor of each pair of numbers.
  a 15 and 20   b 12 and 16   c 6 and 9   d 28 and 42
  e 32 and 36   f 24 and 48   g 40 and 50   h 75 and 80

7 Shawnae has five number cards.
  Her friends each choose a card.
  • Jayson says his card is a factor of 3 and 12.
  • Nisha says her card is not a factor of 12 or 14.
  • Marie says her card is a factor of 18.
  • Marten says his card is a factor of 8.
  Which card is Shawnae left with?

| 3 | 4 | 5 | 6 | 7 |

Looking Back

1 What are the factors of 21?
2 How do you know that 99 is a multiple of 9 but not a multiple of 10?

Unit 2 Prime Numbers and Prime Factors

Let's Think ...

Find all the possible factor pairs of:

7   11   17   23

What do you notice about these numbers and their factors?

Numbers that have more than two factors are called composite numbers.

- 25 is a composite number. Its factors are: 1, 5 and 25.
- 12 is a composite number. Its factors are: 1, 2, 3, 4, 6, 12.

Numbers that have only two factors are called prime numbers.

- A prime number can only be divided by 1 and itself.
- 7 is a prime number. Its factors are 1 and 7.
- 11 is a prime number. Its factors are 1 and 11.
- 1 is not a prime number, nor a composite number, because it only has one factor.

Look at the factors of 12 again:

Factors of 12: 1, 2, 3, 4, 6, 12

There are six factors. Two of them are prime numbers.

- 2 is a prime number because it has two factors: 1 and 2.
- 3 is a prime number because it has two factors: 1 and 3.
- 2 and 3 are the prime factors of 12.

Every composite number can be written as a product of its prime factors.

You can use a factor tree to find the product of primes.

Example 1

Write 12 as a product of its prime factors.

Start with any factor pair.

3 is prime. Circle it.

4 is not prime.

Break 4 into a factor pair.

2 is prime, so circle them both.

All the circled numbers are prime factors.

Write them as a multiplication.

$12 = 2 \times 2 \times 3$

You should write the prime factors in ascending order.

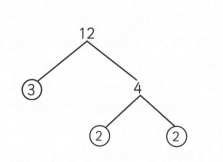

When you use a factor tree you can start with any factor pair. You will still end up with the same prime factors.

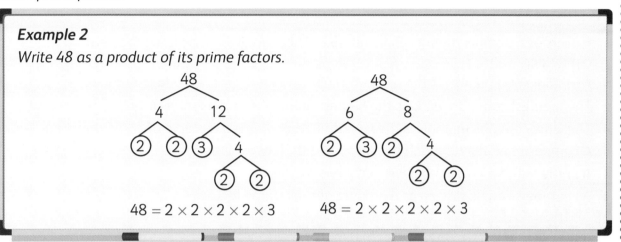

Example 2

Write 48 as a product of its prime factors.

$$48 = 2 \times 2 \times 2 \times 2 \times 3 \qquad 48 = 2 \times 2 \times 2 \times 2 \times 3$$

1 The first five prime numbers are listed here. List the next five.

2, 3, 5, 7, 11

2 The first five composite numbers are listed here. List the next five.

4, 6, 8, 9, 10

3 There is only one even number that is also a prime number.

  a What is the only even prime number?

  b Why are all other even numbers composite numbers?

4 Use the information in the factor trees to write 50 and 96 as a product of their prime factors.

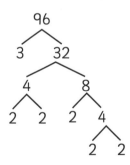

5 Draw your own factor trees to factorize each of these numbers and then write them as a product of their prime factors.

  a 25   b 30   c 18   d 44   e 84   f 64   g 130

Looking Back

1 List the odd prime numbers less than 20.

2 List the composite numbers between 10 and 25.

3 Write 36 as a product of its prime factors.

Topic Review

Talking Mathematics

Make a set of clues for this crossword puzzle.

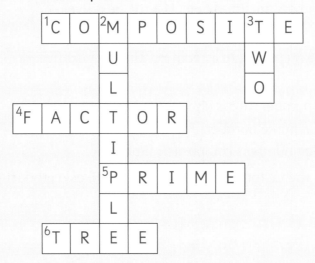

Quick Check

1. a List the factors of 40.
   b List the factors of 48.
   c List all the common factors of 40 and 48.
   d Which of the common factors are prime numbers?

2. a List the first ten multiples of 6.
   b List the first ten multiples of 8.
   c What is the least common multiple of 6 and 8?

3. Here is a set of numbers:
   2, 5, 7, 11, 12, 15, 20, 36, 41
   a List the odd numbers.
   b List the even numbers.
   c List the prime numbers.
   d List the composite numbers.
   e Express each of the composite numbers as a product of its prime factors.

Topic 11 Lines and Angles Workbook pages 33–37

▲ Use the clock above. What angle is made by the minute hand and the second hand? What angle is made by the minute hand and the hour hand? Can you spot any parallel line segments? Why do you think a clock's face is round?

Can you think of any **lines** or angles in nature? What about the horizon?
Look at trees, leaves and spider webs, and see if you can identify **line segments**, different kinds of **angles**, or even **parallel** or **perpendicular lines**.

Lines and angles exist in the world all around us, in the human as well as the natural environment. In geometry, we give very specific definitions to **points**, lines, **planes** and angles. In this topic, you will learn how to draw and name points, lines and planes. You will also learn more about angles and how to measure them.

Getting Started

1 Can you remember the difference between a line, a line segment and a ray?

2 What are parallel lines and perpendicular lines, and can you give any examples?

3 What different types of angles do you know? Can you show them using your arms?

4 What is this instrument called? Discuss with your partner how to use it correctly.

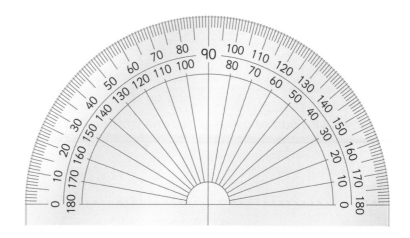

Unit 1 Points, Lines and Planes

Let's Think …

- How would you describe the very tip of your pencil?
- Describe the thin edge of your ruler.
- How would you describe the top surface of your desk?
- What shape is your desk? What can you say about the sides of your desk?

A point is an exact location usually labelled with a capital letter; for example, point P on line AB below. A point has no size, only a position.

Point P

A line is straight and has no endpoints. It has length but no width or height, and therefore it is one-dimensional. You name a line using any two points on the line.

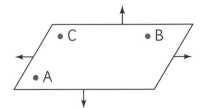

Line EF, Line FE, $\overleftrightarrow{EF}$ or $\overleftrightarrow{FE}$

A plane is a flat, two-dimensional surface that continues forever on all sides. It has length and width but no height. Circles and polygons are examples of plane shapes. To name a plane, you need to include at least three points on the plane that are not on the same line.

Plane ABC, Plane CAB, Plane BCA

Sometimes, you need to work with parts of lines. A line segment is part of a line and has two endpoints.

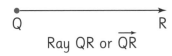

Line segment KL, Line segment LK,
$\overline{KL}$ or $\overline{LK}$

A ray is also part of a line. It has one starting point and the other end continues forever.

Ray QR or $\overrightarrow{QR}$

The relationships between lines are often used in geometry; for example, parallel and perpendicular lines. *Parallel lines* are lines that never intersect but are always the same distance from each other. Think of two sides of a highway that always remain next to each other but never cross. The small arrows on the two lines indicate that the lines are parallel.

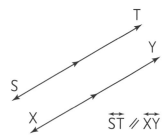

$$\overleftrightarrow{ST} \; / \! / \; \overleftrightarrow{XY}$$

Perpendicular lines do cross or intersect, and always at a right angle. You use a small square to indicates that the angle is 90°. Can you think of any examples of perpendicular lines?

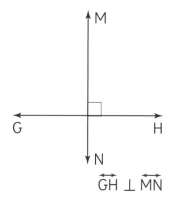

$$\overleftrightarrow{GH} \perp \overleftrightarrow{MN}$$

1 Write down the correct name for each of the following:

a

b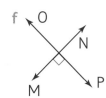

c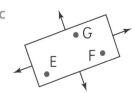

d •H

e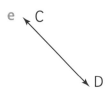

f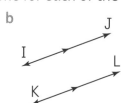

g

2 Answer the following questions.

 a What do you call two lines that cross?

 b What are two lines that cross at a right angle called?

 c What is the name given to two lines that never cross?

 d What is a straight path joining two points called?

 e What do you call a straight path that starts at one point and continues forever in one direction?

 f What does parallel mean?

 g Which lines are perpendicular in the diagram below?

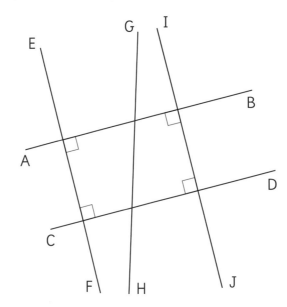

 h Line AB is parallel to line CD and line CD is perpendicular to line EF. What can you conclude about lines AB and EF?

3 Draw and label the following:

 a point B

 b line EF

 c plane RST

 d line segment CD

 e ray GH

 f two parallel lines

 g two perpendicular lines.

Looking Back

With a partner, try to find three real-life examples of each of the geometric definitions you have learned about in this unit.

Unit 2 Angles

Let's Think …

Use two strips of cardboard and a split pin to make an angle strip, almost like the hands of a clock. Make sure that the strips are lined up and your pin is fairly tight but not too tight.

- Use your angle strip to make different sizes of angles.
- What angle do you get when you make the two strips perpendicular to each other?
- Can you make a straight line with your two strips?
- What happens when you keep on rotating the one strip around?

An angle is formed when two rays or line segments meet at a common endpoint called a vertex. Angles indicate the amount of rotation or turning between the two arms, and are measured in degrees.

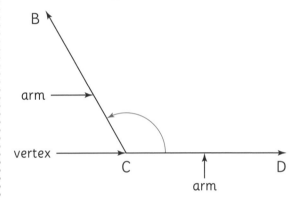

∠BCD or ∠DCB

There are different names for angles depending on how big they are.

- A *right angle* measures exactly 90°. The two arms of the angle are perpendicular to each other.

- A *straight angle* measures 180° and forms a straight line.

- *Acute angles measure less than 90°, so they are smaller than a right angle.*

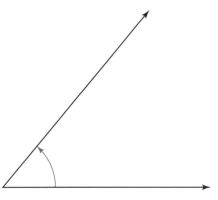

- *Obtuse angles measure more than 90° but less than 180°, so they are in between a right angle and a straight angle.*

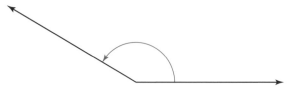

- *When an angle has made a full rotation, the arms meet up again and the angle forms a circle. A full rotation is 360°. When the minute hand of a clock moves from three o'clock to four o'clock, it makes a full rotation from the 12 all the way around back to the 12.*

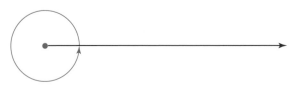

1 Classify each angle as right, straight, acute, obtuse, or full rotation.

a   b   c   d

e   f   g   h

2 Classify each angle as right, straight, acute, obtuse, or full rotation.

a 110°   b 57°   c 92°   d 180°

e 360°   f 88°   g 90°   h 160°

3 Answer the questions that follow.

a Name the vertex and sides of each angle.

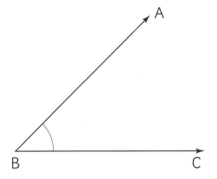

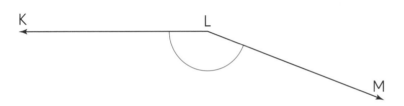

b How many degrees are there in a circle?

c Imagine you are facing north. You turn clockwise by 90 degrees. Which direction are you facing now?

d What does the letter W look like when you rotate it by 180 degrees?

e What angle do the two lines of the letter L make?

4 Try to write your name so that each letter has only right angles in it.

5 List five places in your classroom where you can find examples of right angles.

Looking Back

With your strips of cardboard, show a partner examples of all the types of angles you know. Try to name each angle as your partner makes it.

Unit 3 Measuring Angles

Let's Think ...

- Why can you not use a ruler to measure the size of an angle?
- You use a protractor to measure an angle. Take a closer look at your protractor. What do you notice?
- What numbers and markings are on the protractor?
- Are some lines longer than other lines? Can you tell why?

You use a *protractor* to measure the size of an angle. Look at the protractor shown here.

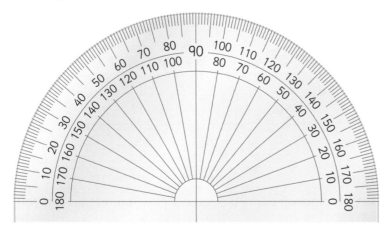

This protractor measures angles up to 180° (a straight angle). If you take two protractors and put the straight edges together so the straight lines overlap, you get a circle. There are 360° in a circle.

Most protractors have two sets of measurement, one on the inside and one on the outside. When you measure an angle, make sure that you always use the scale that starts at zero.

Follow these steps to measure an angle using a protractor.

Step 1: Ensure that the cross at the bottom centre of the protractor is exactly on the vertex of the angle.

Step 2: Line up one arm of the angle with the straight line.

Step 3: Start from 0 and go around the edge of the protractor to where the other arm intersects, and read the number of degrees. Stay on the scale that started at 0!

Check: Check your answer by estimating the size of the angle before you measure it; for example, if you can see that the angle is an obtuse angle, but measure 55°, you know that your measurement is not correct and you have to try again.

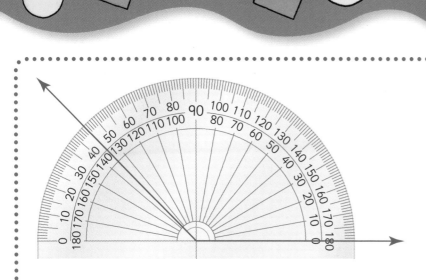

This angle is obtuse and measures 135°.

1 Write down the following protractor readings:

a

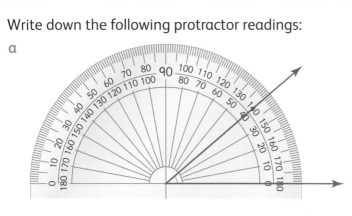

b

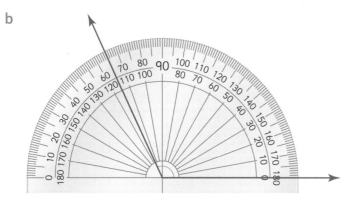

c

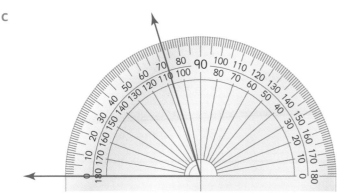

2 Write down the type of angle and use a protractor to measure each angle.

a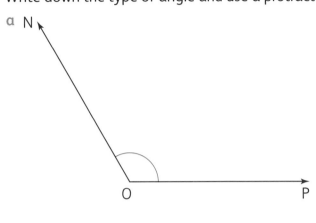

b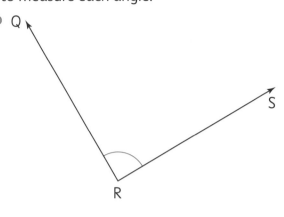

c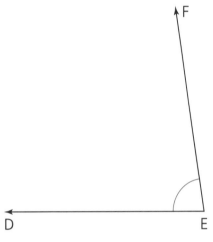

d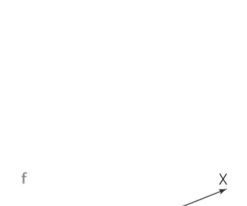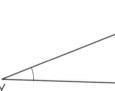

e

f

Looking Back

Draw any two intersecting lines, like those shown here.

Now measure the four angles you have created. Compare your results with three others in your class. What do you notice?

Topic Review

What Did You Learn?

- A point is an exact location with no size, only position.
- A line is straight, continues in both directions and has length but no width or height.
- A plane is a flat, two-dimensional surface that continues forever on all sides with length and width but no height.
- A line segment is part of a line with two endpoints.
- A ray is part of a line. It has one starting point and the other end continues forever.
- Parallel lines never intersect and are always the same distance from each other.
- Perpendicular lines intersect at a 90° (a right angle).
- An angle consists of two arms that meet at the vertex. The size of the angle is measured in degrees.
- A right angle measures 90°, a straight line measures 180°.
- An acute angle measures between 0° and 90°.
- An obtuse angle measures between 90° and 180°.
- A full rotation or circle measures 360°.
- The tool you use to measure the size of an angle is a protractor.

Talking Mathematics

Give the mathematical term for each of the following:

a an angle that is smaller than a right angle

b part of a line with two endpoints

c lines that cross at a 90° angle

d the tool you use to measure angles

e an angle that measures 180°

f the point where the two arms of an angle meet

g something that is straight, one-dimensional and has no endpoints

h an angle that measures 360°

i lines that always remain the same distance from each other

j an angle that is bigger than a right angle but smaller than a straight angle.

Quick Check

Draw an example of each of the following:

- Acute angle
- Ray
- Obtuse angle
- Line segment
- Parallel lines
- Right angle
- Perpendicular lines

Topic 12 Number Facts Workbook pages 38–39

Key Words

fact

recall

fact family

inverse

speed

accuracy

▲ What is the mathematical word for an arrangement of rows and columns like this? Use the display in the photograph to make up one division, one multiplication, one addition and one subtraction number sentence. Compare your sentences with a partner.

How many number **facts** do you think you know from memory? If you know all your basic addition and subtraction facts to 100, then you already know more than 100 facts. If you know your times tables up to 9×9, then you know 100 multiplication facts and you can use these to work out 90 division facts (there are fewer division facts because you can multiply by 0, but you cannot divide by 0). When you know the basic number facts, you can use them to make other calculations much easier and faster.

Getting Started

1 Make up a basic fact quiz with 20 questions. Include addition, subtraction, multiplication and division facts. Write the question and answers on separate sheets.

2 Exchange your quiz with a partner. Complete each other's quizzes in one minute. Your teacher will time you.

3 Hand back the completed quizzes and mark them. Give your partner a rating based on this scale:

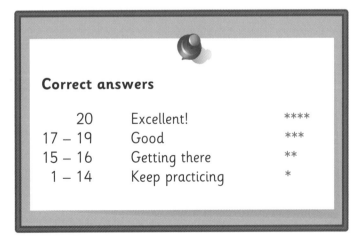

Correct answers

20	Excellent!	****
17 – 19	Good	***
15 – 16	Getting there	**
1 – 14	Keep practicing	*

Unit 1 Know Your Facts

Let's Think ...

Look at these two number sentences. $19 + 8 - 8 = 19$ $7 \times 8 \div 8 = 7$

Explain why the answer is the same as the starting number in each calculation.
Write two more examples like these.

The four basic operations are addition, subtraction, multiplication and division.

The four operations are related to each other so you can use facts that you know to work out other related facts.

Look at the diagram and read the examples. Think about how this can help you work out and remember facts.

$$12 + 3 = 15$$
$$15 - 3 = 12$$
$$15 - 12 = 3$$

 add 4 times

| Addition + | ← inverse → (undo each other) | Subtraction − | 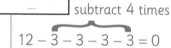 subtract 4 times |

$$3 + 3 + 3 + 3 = 12$$
$$3 \times 4 = 12$$

$$12 - 3 - 3 - 3 - 3 = 0$$
$$12 \div 3 = 4$$

You are expected to know the basic facts and to be able to recall them quickly and correctly. The key to improving your speed and your accuracy is to keep practicing and using the facts.

| Repeated | | Repeated |

| Multiplication × | ← inverse → (undo each other) | Division ÷ |

$$5 \times 7 = 30$$
$$30 \div 5 = 7$$
$$30 \div 7 = 5$$

---

1 Write the answers only. Try to do each set in under 90 seconds.

Set 1

a 3×5	b 5×8	c 8×0	d 8×10
e 10×4	f 6×8	g 5×6	h 4×9
i $12 \div 2$	j 4×4	k $15 \div 5$	l $32 \div 8$
m 7×3	n 6×2	o 0×9	p $20 \div 4$
q $54 \div 6$	r $24 \div 3$	s $20 \div 2$	t $49 \div 7$

Set 2

a $12 + 8$	b $14 - 4$	c $19 - 9$	d $12 - 12$
e $18 + 3$	f $9 + 9$	g $12 - 6$	h $18 - 9$
i $8 + 14$	j $6 + 9$	k $4 + 6$	l $5 + 7$
m $20 - 8$	n $30 - 15$	o $12 - 10$	p $21 - 21$
q $19 - 3$	r $17 - 9$	s $17 - 7$	t $3 + 19$

2 Use number facts to answer these questions without doing any working out.

   a How many days in 4 weeks?

   b What is half of 22?

   c How many sides are there in 8 pentagons?

   d What is the change from $20.00 if you spend $5.00?

   e How many sweets would each person get if you shared 24 sweets equally among four people?

   f How many groups of 5 can you make from 40 stickers?

   g If there 9 seats per row, how many seats in 8 rows?

   h What is the perimeter of a square with sides 6 cm long?

   i If you cut 9 cm from a 30 cm strip, what is the length of the leftover strip?

   j Josh takes 19 minutes to get to school. Maria takes 12 minutes. What is the difference between the times?

3 Look at the shapes. Answer the questions using number facts.

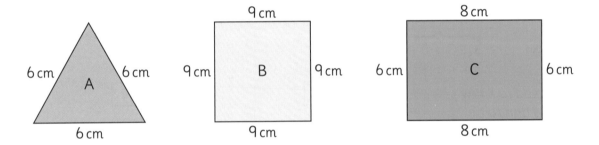

   a What is the perimeter of shape A?

   b What is the perimeter of shape B?

   c What is the area of shape C?

   d What is the area of shape B?

   e What is the combined length of the two longest sides of shape C?

   f What is the combined length of the two shortest sides of shape C?

Looking Back

1 Write the answers only. Time yourself.

a 11 − 3	b 9 + 6	c 4 × 9	d half of 20
e 18 ÷ 2	f 6 + 8	g 6 × 7	h double 9
i 11 − 9	j 2 + 16	k 9 × 8	l 16 − 11
m 5 + 4	n 5 × 4	o 28 ÷ 4	p 40 ÷ 8

2 Sondra looks at her watch at 9 minutes past 2. She looks at it again 8 minutes later. What time will it be?

3 A regular pentagon has sides that are 7 cm long. What is the total distance round the pentagon?

Topic Review

What Did You Learn?

- You did not learn anything new in this topic. You practiced the number facts you have already learned to improve your speed and accuracy.

Talking Mathematics

1 Think about the multiplication and division fact family for 4, 9 and 36.
- Which numbers are factors?
- What number is a product?
- Which numbers can be used as divisors?
- Which numbers are quotients?

2 An addition and subtraction fact family has 19 as the greatest number and 7 as one of the other numbers.
- What is the third number in the family?
- What related facts can you make for this family?

Quick Check

1 Write three related facts for each number sentence.

a $15 - 8 = 7$
b $25 - 9 = 14$
c $23 - 6 = 17$
d $9 + 8 = 17$

e $32 - 8 = 15$
f $18 - 13 = 5$
g $24 \div 8 = 3$
h $40 \div 5 = 8$

i $56 \div 8 = 7$
j $9 \times 6 = 54$
k $3 \times 9 = 27$
l $6 \times 8 = 48$

2 Fact check. Work with a partner. Take turns to read a fact. Say whether it is TRUE or FALSE. Check with a calculator. Score 1 point for every correct answer.

a 18 is a multiple of 3
b 9 groups of $4 = 54$
c 60 is a multiple of 10
d 12 is a factor of 6
e 7 multiplied by 4 is 21
f 1 multiplied by itself is 1
g 15 taken away from 15 is 0
h the product of 4 and 5 is 20
i the sum of 4 and 5 is 20
j the difference between 17 and 9 is 8
k the sum of 7 and 8 is 16
l double 9 is 18
m 30 divided by 6 is 5
n 45 divided by 9 is 5
o 14 less 9 is 5
p the difference between 17 and 12 is 5
q 81 divided by 9 is 9
r 29 is a multiple of 2
s 12 is a factor of 24
t 8 is a factor of 32

3 a A square has a perimeter of 25 cm. How long is each side?

b A rectangle has a perimeter of 30 cm. If one side is 10 cm long, how long are the other three sides?

c What is the area of a rectangle with sides of 6 cm and 9 cm?

Topic 13 Adding and Subtracting Workbook pages 40–41

▲ The smaller cruise ship carries 2 744 passengers and has 833 crew members. The larger ship carries 5 412 passengers and has 2 394 crew members. How would you work out the total number of passengers carried by both ships? How would you work out how many more crew the larger ship needs? How would you estimate the difference in passenger numbers?

You already know how to **estimate** and how to **add** and **subtract** large numbers using different methods. You should also be familiar with the mathematical language that tells you whether to add or subtract. You will practice your skills in this topic and apply them to solve one-step and multi-step problems in which you add and/or subtract to find solutions.

Getting Started

1 Use the figures above to answer the following questions.

  a What is the number of crew altogether on both ships?

  b How many more passengers does the large ship carry than the smaller one?

  c How many fewer crew than passengers are there on the small ship?

  d How many passengers and crew are there in total on each ship?

2 a Make up two simple problems of your own using the information given about the cruise ships and solve them.

  b Use a calculator to check if your solutions are correct.

  c Exchange problems with a partner and solve each other's problems.

  d Check your partner's solution and explain where he or she went wrong if they made any mistakes.

Unit 1 Revisit Addition of Larger Numbers

Let's Think …

On one weekend, 10654 cruise passengers visited an island on Saturday and 9875 visited on Sunday. Estimate the total number of cruise passengers for the weekend. Tell your partner how you found your answer.

Do you remember how to *add* numbers using expanded notation and the column method?
Read through the examples to remind you how to use each method.

Example 1 Add 32452 + 12303

Estimate 32000 + 12000 = 44000

32452	30000 + 2000 + 400 + 50 + 2	Use expanded notation.
12303	10000 + 2000 + 300 + 3	Line up the places.
	40000 + 4000 + 700 + 50 + 5	
	= 44755	Write the answer in standard form.

Example 2 Find the sum of 54214 + 3751

Estimate 54000 + 4000 = 58000

```
  54214      Write digits with the same place value below each other.
+  3751      Add up the columns.
  57965
```

Example 3 What is 9769 + 1489?

Estimate 10000 + 1000 = 11000

```
  1 1 1
   9769      Line up digits with the same place value.
+  1489      Add up the columns.
  11258      Regroup as you need to. Carry the tens to the next place.
```

9 + 9 = 18, write 8 in the ones place and carry 1 to the tens place.

6 + 8 = 14, 14 + 1 (regroup) = 15, write 5 in the tens place and carry 1 to the hundreds place.

7 + 4 = 11, 11 + 1 (regroup) = 12, write the 2 in the hundreds place and carry 1 to the thousands place.

9 + 1 = 10, 10 + 1 (regroup) = 11. There are no more digits to add, so write 11 (thousands) in the answer line.

1   Estimate by rounding and then add.
   a 455 + 223
   b 160 + 725
   c 2 345 + 1 054
   d 4 281 + 1 110
   e 6 330 + 1 234
   f 342 + 11 013
   g 11 321 + 33 107
   h 423 + 11 024
   i 105 + 89 235
   j 129 300 + 489
   k 40 + 125 608
   l 562 + 987 417

2   Write in columns and add.
   a 12 + 60 + 13
   b 123 + 23 + 4 000
   c 241 + 2 400 + 10 000
   d 12 400 + 238 + 200 000
   e 32 + 400 + 32 000 + 12
   f 50 + 500 + 50 200

3   Add using the method you find easiest. Write an estimate before you start working.
   a 437 + 876
   b 398 + 1 209
   c 5 427 + 2 686
   d 12 987 + 4 567
   e 5 412 + 19 234
   f 28 435 + 32 876
   g 178 796 + 54 321
   h 123 987 + 12 450
   i 124 999 + 654 230
   j 345 076 + 32 987

4   Estimate and then calculate the answer.
   a What is the sum of 12 345 and 14 098?
   b Add 129 452 to 897 806.
   c Find the sum of 1 345, 12 304 and 13 098.
   d What is the total of 45 678 and 890 121?
   e What number is 13 454 greater than 13 454?

5   Write down three real-life situations where an estimated total would be good enough.

6   Write down three real-life situations where an estimated total would not be good enough and could cause problems.

Looking Back

Calculate:
a 26 + 6 543 + 8 + 405
b 4 309 + 19 + 32 456 + 5
c 3 243 + 681
d 233 793 + 26 134

Unit 2 Revisit Subtraction of Larger Numbers

Let's Think …

A cruise ship carrying 5 409 passengers docked in Nassau. 4 023 passengers left the ship.
About how many remained on board?
Tell your partner how you worked out your answer.

You can use the same methods that you used for addition to *subtract* larger numbers.
Read through the examples below to remind yourself how to subtract with and without renaming numbers.

Example 1   What is 187 324 − 123 103?

Estimate   190 000 − 120 000 = 70 000

187 324   100 000 + 80 000 + 7 000 + 300 + 20 + 4
123 103   100 000 + 20 000 + 3 000 + 100   + 3
                    60 000 + 4 000 + 200 + 20 + 1   Subtract the values in each place.
                        = 64 221

Example 2   Subtract 12 345 from 295 865.

Estimate   300 000 − 10 000 = 290 000

  295 865
− 12 345
  283 520

Example 3   Calculate 14 675 − 12 399.

Estimate   15 000 − 12 000 = 3 000

Subtract the ones, rename the 7 to get 6 tens and ten ones.
15 − 9 = 6

$$\begin{array}{r} 1\ 4\ \overset{5}{\cancel{6}}\ \overset{16}{\cancel{7}}\ 5 \\ -\ 1\ 2\ 3\ 9\ 9 \\ \hline 2\ 2\ 7\ 6 \end{array}$$

Subtract the tens, rename the 6.
16 − 9 = 7

1 Estimate and then subtract to find the answer.

a 689 – 325
b 827 – 304
c 999 – 888
d 1 400 – 1 200
e 3 225 – 224
f 8 234 – 4 317
g 24 245 – 19 321
h 313 098 – 200 450
i 342 512 – 124 755
j 876 312 – 35 980
k 899 765 – 12 098
l 748 982 – 345 897

2 The population of each of the six largest cities in the USA (in 2016) is given below.

| Philadelphia 1 463 281 | Chicago 2 842 518 | New York 8 143 197 |

| Houston 2 016 582 | Phoenix 1 461 575 | Los Angeles 3 844 829 |

a List the names of the cities and their populations in descending order (greatest population first).

b How many more people live in New York than in Los Angeles?

c What is the difference in population between Philadelphia and Phoenix?

d The total population of two of the cities is 4 859 100. If one of the cities is Houston, which is the other?

e How many more people would need to move to Phoenix for it to have the same population as Chicago?

3 Work with a partner.

a Make up four subtraction problems of your own using the city populations. Work out the answers on a separate sheet.

b Exchange your problems with another pair and solve each other's problems.

c Swap back and check each other's work.

Looking Back

Work out the missing value in each of these number sentences.

a 3 816 581 – ☐ = 978 657

b ☐ – 485 728 = 925 245

Unit 3 Mixed Problems

Let's Think …

- Work with a partner. Read through this problem.
- Talk about how you would solve it.
- How many calculations would you need to do?
- How much greater is the sum of 15 308 and 8 745 than the difference between 321 456 and 189 450?
- Once you have talked about it, work out the solution.

You can use addition and subtraction to solve many different kinds of mathematical problems. But, before you do any calculations, you need to read the problem and decide which operation or operations you need to do.

There are often clues in the word problems that tell you whether to add or subtract.

Some clues that you need to add:
How far …
How many …
How much …
Sum
… in all
… total
… altogether

Some clues that you need to subtract:
How much further …
How many more/less …
How much more/less …
How much is left …
Difference
Greater than …
Less than …
… more than

Sometimes you need to do more than one operation to solve a problem. Read through the example to see how two students solved the same problem in different ways. Both did two steps to get to the answer.

Example

The total population of an island is 23 630. If there are 7 562 adult men and 7 542 adult women, how many children are there?

James

$$\begin{array}{r} {}^{1}2\,{}^{1}3\,{}^{5}6\,{}^{1}3\,{}^{2}0\,{}^{1} \\ -\ \ 7\,5\,6\,2 \\ \hline 1\,6\,0\,6\,8 \end{array}$$ Subtract the men.

$$\begin{array}{r} 1\,{}^{5}6\,{}^{1}0\,6\,8 \\ -\ \ 7\,5\,4\,2 \\ \hline 8\,5\,2\,6 \end{array}$$ Subtract the women.

There are 8 526 children.

Amalie

$$\begin{array}{r} {}^{1}7\,{}^{1}5\,6\,2 \\ +\ \ 7\,5\,6\,2 \\ \hline 1\,5\,1\,0\,4 \end{array}$$ adults

$$\begin{array}{r} {}^{1}2\,{}^{1}3\,6\,{}^{2}3\,{}^{1}0 \\ -\,1\,5\,1\,0\,4 \\ \hline 8\,5\,2\,6 \end{array}$$ children

There are 8 526 children.

Before you try these problems, read them carefully and work out what operation you need and whether you need to do more than one calculation.

1   What is 237 less than the sum of 13 453 and 8 765?

2   A car salesperson buys a car for $23 456.00 and sells it on for $25 000.00. How much money did he make on the sale?

3   The coastline of The Bahamas is estimated to be 11 238 km long. The coastline of Cuba is estimated to be 14 519 km long.
    a   What is the length of the combined coastline of the two countries?
    b   How much greater is the coastline of Cuba than that of The Bahamas?

4   An estate agent buys a property for $127 800.00. She spends $1 500.00 to have it divided into two plots. She then sells one plot for $80 275.00 and the other for $65 225.00. Work out how much money she makes on the deal.

5   A farmer had 13 753 banana trees on one plantation and 18 348 on another. During a hurricane, 2 367 trees were destroyed. How many were left?

6   At the beginning of a year, Mrs Smith had $26 732.00 in her savings account. At the end of the year, she had $28 617.00 in the account. How much more did she have at the end of the year?

7   A bookshop had 125 000 books in stock. 19 250 were sold in one month and 87 750 were sold in the next three months. How many were left in the stock after four months?

8   The solutions to three different problems are given below. For each one, write a word problem that fits the number sentences.
    a   3 456 + 12 876 = 16 332
    b   763 + 987 = 1 750
        1 750 − 1 321 = 429
    c   28 + 147 + 305 = 480
        756 − 480 = 276

Looking Back
A cruise ship leaves Fort Lauderdale with 2 309 passengers on board. It stops in Miami where another 1 306 passengers get on board. When it gets to Grand Bahama 298 passengers leave the cruise to fly home and another 407 board. How many passengers are on board for the return cruise?

Topic Review

Talking Mathematics

Work in pairs. Pretend you are teachers.

Make a poster for your class to teach them how to read and tackle word problems involving more than one step. Use examples if you need to.

Quick Check

1 Write each of these numbers in numerals.

   a three thousand four hundred thirty-seven

   b two hundred twenty-one thousand eight hundred twenty-nine

   c twelve thousand eight hundred forty-seven

   d four hundred twenty-three thousand four hundred two

2 Use the numbers in question 1.

   a What is the sum of a and c?

   b What is the total of all four numbers?

   c Find the difference between the sum of a and b and the sum of c and d.

3 Calculate.

   a $10\,991 + 234 + 4\,568$

   b $132\,819 + 343\,214$

   c $112\,345 + 123\,145$

   d $11\,285 - 9\,873$

   e $29\,876 - 14\,388$

   f $234\,000 - 39\,453$

4 Round to the nearest ten thousand and estimate the sum of 11 270, 22 701, 54 688 and 13 431. Calculate the difference between your estimate and the actual total.

5 The sum of three numbers is 12 345. Write four different addition sums that will give this result.

6 The difference between two six-digit numbers is 2 876. What could the numbers be?

Topic 14 Statistics Workbook pages 42–45

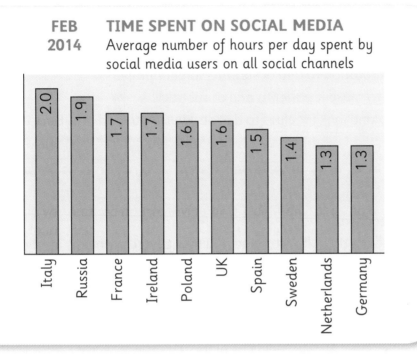

FEB 2014

TIME SPENT ON SOCIAL MEDIA
Average number of hours per day spent by social media users on all social channels

| Italy | Russia | France | Ireland | Poland | UK | Spain | Sweden | Netherlands | Germany |
| 2.0 | 1.9 | 1.7 | 1.7 | 1.6 | 1.6 | 1.5 | 1.4 | 1.3 | 1.3 |

Key Words
mean
median
mode
range
table
line graph
bar graph
double bar graph
Venn diagram

▲ The bar graph above shows the average number of hours spent per day by people who use social media in a few European countries. In which countries would you choose to advertise a product on social media? Why?

Statistics are widely used today in fields as diverse as medicine, politics, insurance and consumer studies. You start the statistical process by collecting data. You can then organize and visually represent the data using a **table** or a **graph.** After that, you can begin asking questions based on the data. This means that you interpret and analyse the data.

In this topic, you will revisit the **arithmetic mean, median, mode** and **range.** You will also look at making decisions about how to represent and analyse data. Finally, you will learn about two more ways of representing data: a **double bar graph** and a **Venn diagram.**

Getting Started

- Can you name three ways of collecting data?
- Name three types of data displays.
- Can you explain what the mean, median and mode of a data set are?
- Explain what the range of a data set is.

Unit 1 Mean, Median, Mode and Range

Month of Year	Jan	Feb	Mar	Apr	May	Jun	Jul	Aug	Sep	Oct	Nov	Dec
Average Air Temperature (°C)	24	24	25	26	28	30	31	31	30	29	26	25

The table above shows the average temperature per month at Nassau Airport.

- Is there a big difference between the averages of the hottest and the coolest months?
- Are there any months that have the same average temperature?
- What would you advise tourists who love hot weather?

The *mean* of a set of data is the number you get when you add all the values in the set and then divide that sum by the number of values in the set.

Find the mean of 2, 7, 5, 4, 6, 3

$2 + 7 + 5 + 4 + 6 + 3 = 27$. There are 6 numbers, so the mean is

$$\frac{27}{6} = 4\frac{3}{6} = 4.5$$

You can see that the mean is not always a whole number.

The *median* is the middle number of an ordered set of data.

2, 3, 4, 5, 6, 7

Since there are an even number of values, there are two values exactly in the middle: 4 and 5. So you take the mean of the two numbers.

$$\frac{4+5}{2} = \frac{9}{2} = 4.5$$

In this instance, the mean and the median are the same. However, this is not always the case.

The *mode* of a set of data is the value that occurs most often. In the data set below, the mode is 4.

1, 2, 3, 4, 4, 4, 4, 4, 5, 5, 5, 6, 6, 7

The *range* is the difference between the greatest and the least values. The range tells you how widely the data set is spread. In the data above, the greatest value is 7 and the least value is 1. So the range is given by:

$7 - 1 = 6$

1 In his end of year exam, Stephen had the following scores:

English 55 %
History 34 %
Geography 34 %
Mathematics 90 %
Science 87 %

 a Calculate his year-end average.
 b Find the median.
 c Work out the mode.
 d What is the range of his scores?

2 The two tables below show monthly rainfall data for Disney World in Florida, USA, and Tokyo Disneyland in Japan.

Disney World, USA: Monthly Rainfall in mm

Jan	Feb	Mar	Apr	May	Jun	Jul	Aug	Sep	Oct	Nov	Dec
59	71	77	51	76	175	208	211	162	66	43	55

Tokyo Disneyland, Japan: Monthly Rainfall in mm

Jan	Feb	Mar	Apr	May	Jun	Jul	Aug	Sep	Oct	Nov	Dec
50	72	106	129	144	176	136	149	216	194	96	54

 a What is the mean monthly rainfall for the year in Tokyo Disneyland?
 b What is the mean monthly rainfall for the year in Disney World?
 c Which park has the highest rainfall on average?
 d What is the mean monthly rainfall from June to August for Disney World?
 e What is the mean monthly rainfall from June to August for Tokyo Disneyland?
 f Which park is the wettest from June to August?

3 A sports store is offering a discount on a particular type of running shoe. The manager wants to look at the number of each size sold so far in order to predict how many will be sold in the next week. She can then make sure that there are enough pairs of each size in the store. Here is a list of all the shoes sold per size:

7, 8, 8, 8, 8, 9, 9, 10, 10, 11, 11

 a Calculate the mean, mode and median of the data.
 b Which of the three measures is most useful to the manager?

> **Looking Back**
> Are you now able to calculate the mean, median, mode and range for the data on the average temperatures per month at Nassau Airport, given on page 87? Write down each of these measures.

Unit 2 Analysing and Representing Data

Let's Think ...

- What kinds of graphs have you learned about and drawn?
- What kinds of data are these graphs suitable for?

A table can be used to organize data.

You can use a graph to display data in a format that makes it easy to see the important information.

Before you create a graph, you must decide which type of graph is most appropriate for your set of data. You need to think about the purpose of your graph and what you want it to show.

Line graphs can show changes over time. They allow you to see overall trends such as an increase or decrease in data over time.

This line graph shows the level water in a reservoir for the month of August. You can see how the water level changed over time. At first, the water level remained more or less constant, then there is a downward trend until the water level is 0, which means the reservoir is empty.

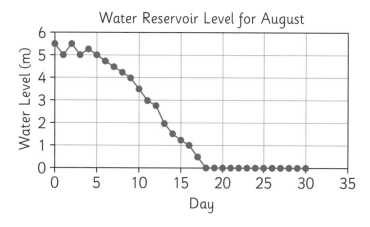

What other observations can you make about the data shown?

Bar graphs are used to compare data. The bars let you compare numbers in different categories and help you to see relationships between the data. Bar graphs can be vertical or horizontal.

This bar graph clearly shows you how much each type of plant has grown in one month. You can quickly see which plant has grown the tallest. You can also see that the pea and the tomato have grown the same amount.

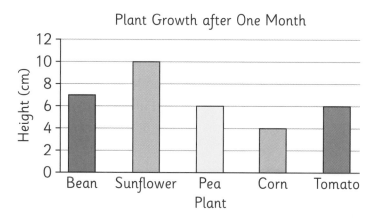

What else can you tell from the graph?

1 Which type of graph would you use to best illustrate the following:

   a the number of books checked out at the school library per month for the past year

   b the number of votes for each candidate running for president

   c the number of internet users in 10 different countries

   d the amount of sugar in different types of food

   e the number of teenagers with cell phones in your country over the past 10 years.

2 Think back to the line graph showing the water level of a reservoir in this unit. Each of the graphs below show the water level of a different reservoir for the same month. What can you tell about the water level of each reservoir as the month progressed?

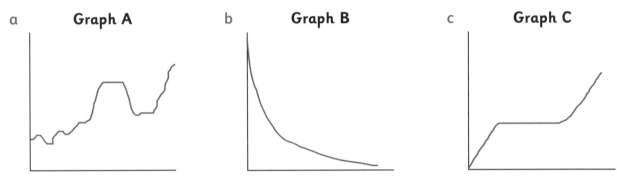

   a   **Graph A**   b   **Graph B**   c   **Graph C**

3 The line graph below shows Jan's electricity charges for the first six months of the year. Study the graph, then answer TRUE or FALSE to each of the statements.

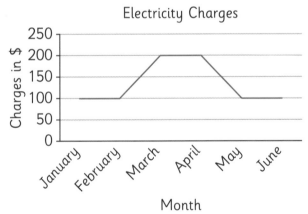

   a The electricity charges went up after April.

   b The charges were stable between March and April.

   c The charges were at their lowest after March.

   d The electricity charges varied by $100.00 only.

Looking Back

List the differences and similarities between bar graphs and line graphs.

Unit 3 More Visual Representations of Data

Let's Think …

● How does a bar graph show information? What does it mean when one bar is longer than another?

● Have you ever heard of a Venn diagram? If so, what do you know about it?

A teacher did a survey in class about students' favourite sport.

The bar graph on the right shows the results. Which sport is the most popular?

The teacher decides that the bar graph does not give an accurate enough picture of his students' interest in sports, as the data for the boys and girls is quite different. So he decides to use a *double bar graph* to show the data.

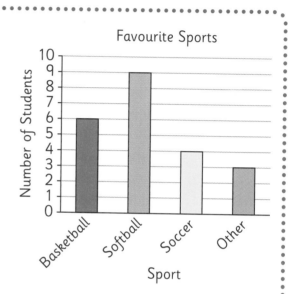

You can see from the graph that softball is the favourite sport among the girls but not the boys. They boys' favourite sport is basketball.

A *double bar graph* displays information about two related sets of data. It shows the data for two different groups, for example girls and boys, but with the same categories. This allows you to compare the data for the different groups.

The bars can be vertical or horizontal.

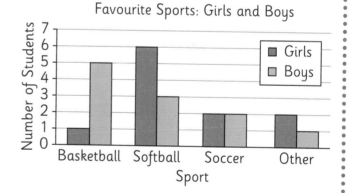

The next day, the teacher did a survey which asked the students whether they had a pet dog or cat. He found that some students had a dog, some had a cat, some had both and some had neither.

91

The best way to represent this information visually, is in a *Venn diagram.*

The Venn diagram shows that three students, Tyrone, Sue and Sonya have only a cat. Steven and Harry have only a dog.

Kailyn, Bernadette and Terrence have both a cat and a dog. Since these three students are included in both sets, their names are placed where the two circles overlap or intersect.

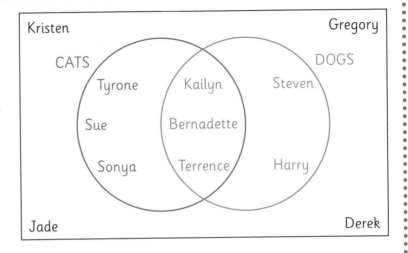

The names of the students who do not own any cats or dogs, are placed outside of both circles, for example Kristen.

In this Venn diagram, there are two overlapping sets:

CATS = {Bernadette, Kailyn, Sonya, Sue, Terrence, Tyrone} and

DOGS = {Bernadette, Harry, Kailyn, Steven, Terrence}

In mathematical notation, Sue $\in$ CATS means that 'Sue' is an element, or member, of the set CATS.

CATS $\cup$ DOGS is the union of both sets and includes every element in both sets.

CATS $\cup$ DOGS = {Bernadette, Harry, Kailyn, Sonya, Steven, Sue, Terrence, Tyrone}

CATS $\cap$ DOGS is the intersection of both sets which includes only the elements that overlap.

CATS $\cap$ DOGS = {Bernadette, Kailyn, Terrence}

1 Lucy is having a party and did a survey among her friends and family to find out what kind of music they like. The results are shown in the bar graph below.

Preferred Music: Adults and Children

a What is the favourite and the least favourite music among the children?

b What is the favourite and the least favourite music among the adults?

c What would be the best music to choose to ensure that both adults and children have a good time?

2 A mathematics teacher surprises his class with an unprepared test. He then announces that the class will write another test in a week's time. The results of the two tests are shown below.

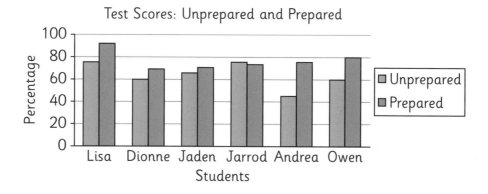

Test Scores: Unprepared and Prepared

Write down ten observations you can make from the graph.

3 The Venn diagram shows the courses that students are taking in art class.

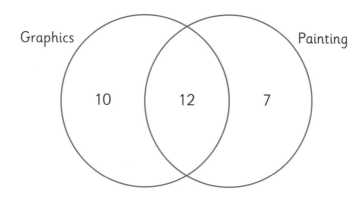

a How many students are taking graphics and painting?
b How many students are taking graphics only?
c How many students are taking painting?
d How many students are there in art class altogether?

4 You have two sets of numbers, C = {1, 3, 5, 7, 9} and D = {2, 3, 5, 7}.
a TRUE or FALSE: 3 ∈ C?
b TRUE or FALSE: 1 ∈ D?
c What is C ∪ D?
d What is C ∩ D?
e What is in set C but not in set D?
f Write down one number that is outside both sets.

Looking Back
Write a note to a friend in which you explain what a double bar graph and a Venn diagram are. Use your own words to explain these two visual representations to your friend.

Topic Review

What Did You Learn?

- The mean or average of a set of data is the sum of all the values in the set divided by the number of values in the set.
- The median is the middle number in an ordered set of data.
- The mode of a data set is the value that occurs most often.
- The range of a data set is the difference between the greatest and the least values.
- Graphs are used to give a visual picture of collected data. It is important to consider the purpose of the graph and the information that needs to be communicated before choosing the type of graph.
- Tables are used to display and organize data.
- A line graph is a useful way to show changes over time and to see trends and patterns.
- A bar graph is useful for comparing data as it shows the numbers in different categories.
- A double bar graph is a bar graph that shows information about two related sets of data, allowing you to compare the data for the two different groups.
- A Venn diagram shows the relationships between two overlapping sets of data.

Talking Mathematics

What is the mathematical word for each of these?

- $\in$
- The value in a set of data that occurs most often.
- A graph that uses rectangular bars to show how big each value is.
- The middle number of an ordered set of data.
- A graph that shows information for two related sets of data in bars.
- A graph that shows changes over a period of time.
- The value that tells you how widely a data set is spread.

Quick Check

1 What do you need to do before you can determine the median of a set of data?
2 Why can you spot trends from a line graph?
3 How do you calculate the mean of a set of data?
4 Why would someone use a double bar graph instead of a single bar graph?
5 What is included in the intersection of two sets?

Topic 15 Problem Solving Workbook pages 46–48

Key Words
pattern
strategy
non-routine
operations
solve
equation
expression
unknown
variable

▲ This is a combination lock padlock. Each round section has all the digits from 0–9 on it. You choose three as your passcode and only those three in the correct order will unlock the padlock. How many different passcodes do you think you can you make if you do not repeat any digits? Think carefully about how you could work this out. Discuss your ideas with your group.

When you have to solve problems in mathematics, it helps to work systematically. First, you need to read and understand the problem, then you have to choose a suitable strategy to **solve** it. Next, you work out the solution and lastly you check that it seems reasonable and correct.

Some of the different strategies you have already used are drawing diagrams, guessing and checking, finding a **pattern**, making a list or table or writing a number sentence. Writing number sentences to represent problems is a very useful strategy. In this topic, you are going to learn how to write a number sentence to show a problem when you do not know some of the numbers involved. As you work through this topic, you will try different strategies to solve multi-step and **non-routine** problems.

Getting Started

1 Jania sets a code on a padlock like the one above. She writes these clues down so she can remember what the password is. Can you work out what it is?

2 James says he is thinking of a number. If he divides his number by 60, he gets 50. What number is he thinking of?

> It has three different odd digits.
> If it was a 2-digit decimal, it would round to 5.8.
> It is divisible by 3 but not by 9.

Unit 1 Problem Solving Strategies

There are different ways to solve problems.

You need to choose the strategy that best fits the problem you are trying to solve.

Strategy	When Is It Useful?	Tips
Trial and error: also called 'guess, check and refine'	You are not sure of the answer, but have a 'feeling' about what it might be.	Do not make wild guesses. Use the information you have to guess sensibly. Check if the answer is correct. If not, use what you have learned from your guess to get closer to the correct answer.
Sketch a drawing, diagram or model	The information is easier to sketch than to write out. These problems may be shape or measurement problems.	Label your sketch properly with all the necessary information.
Make a list, chart, table or tally	You are working with different combinations, probabilities or sets of data.	Choose the option that best matches your information.
Eliminate possibilities	There are only a few possible correct answers. Eliminate those that do not work.	Use tables or grids. Cross off answers as you eliminate them.
Use inverse operations	The problem calculation uses one of the four operations. However, some information is missing.	Use the information you already have to set up the inverse operations.
Write a number sentence (equation).	The problem gives numerical information and some clues about what operations you would need to solve it.	Always set out your working clearly and show any working out that you do.

Work in pairs or small groups. Decide what strategy is most useful for solving each problem and then work on your own to solve it.

1 The numerals 1 to 7 are written on seven different cards.

   a How many pairs of cards add up to 8?

   b How many ways are there to group three cards and get a total of 10?

2 A large boat-building company makes 15 dinghies per day, five days per week for 48 weeks of the year. How many dinghies do they make in a year?

3 Jess makes this grid of squares using matchsticks.

She puts one stone in each square. She then removes one stone and rearranges the matchsticks so that each of the three remaining stones is still inside a square. Sketch the new arrangement.

4 Mr Nixon has 2 lengths of rope. One is 1.895 m long, the other is 1.904 m long.

   a Which piece is longer and by how much?

   b If he needs 4.5 metres of rope in all, how much more will he need to buy?

5 Look at this puzzle and its solution. Each letter represents a single-digit number. If the letter appears more than once, it has the same value each time it appears.

$$
\begin{array}{rr}
\text{FORTY} & 29\,786 \\
\text{TEN} & 850 \\
+\ \ \text{TEN} & 850 \\
\hline
\text{SIXTY} & 31\,486 \\
\end{array}
$$

Here are some examples. Try to find a set of numbers that works for each one.

a
$$
\begin{array}{r}
A \\
B \\
+\ AB \\
\hline
BA \\
\end{array}
$$

b
$$
\begin{array}{r}
CCC \\
+\ \ D \\
\hline
DEEE \\
\end{array}
$$

c
$$
\begin{array}{r}
M \\
M \\
+\ M \\
\hline
LM \\
\end{array}
$$

Now try these. Find at least two different number solutions for each calculation.

d
$$
\begin{array}{r}
\text{NINE} \\
-\ \text{FOUR} \\
\hline
\text{FIVE} \\
\end{array}
$$

e
$$
\begin{array}{r}
\text{THREE} \\
+\ \text{FOUR} \\
\hline
\text{SEVEN} \\
\end{array}
$$

Looking Back

There are five boys in a room. Andrew, Byron, Colin, David and Ethan.

The teacher wants three of the boys to help her.

How many different options does she have to choose three boys?

Unit 2 Expressions and Variables

Let's Think …

What do the following mean in mathematics?

- 2 more than a number
- 3 less than a number
- 3 times as much as a number
- the product of a number and 4

How could you write these using operation symbols? Share your ideas with your group.

In mathematics, we often have to *solve* problems involving *unknown* amounts; for example, the sum of two different whole numbers is 11. What could the numbers be?

In this example, you do not know what the numbers are, but you do know that if you add them together you get 11.

In lower grades, you may have written this as $\square + \bigcirc = 11$

The different shapes in the number sentence show that the two numbers are different.

You can use letters rather than empty shapes to stand in for unknown numbers in mathematics.

You could write this problem as $a + b = 11$

In this example a and b stand for different numbers, but the value of a and b can vary:

if a is 1, then b must be 10

if a is 2, then b must be 9

Because the value can change (or vary), we call these letters *variables*.

The number sentence $a + b = 11$ is called an *equation*.

Any number sentence with an equals sign is an equation.

Sometimes you need to represent a situation without using an equals sign.

When you join variables and numbers with operation symbols (but not an equals sign), it is an *expression*.

Here are some examples of expressions:

- *The sum of two unknown numbers* $a + b$
- *Twice the value of an unknown number* $2 \times m$
- *2 less than an unknown number* $n - 2$
- *A number divided by 3* $x \div 3$
- *3 more than twice a number* $3 + 2 \times y$

You can use any letters to represent unknown numbers, but you should say what they represent.

Example

Melissa has some counters in her pocket. She loses 5 of them. How many are left.

Let the number of counters be c.

She has c − 5 counters left.

It is a rule that we use lower-case letters rather than capital letters.

1 Work in pairs. Discuss what these expressions mean.

a $x + 6$ b $m - 2$ c $4 \times y$ d $40 + x$

e $3 \times m + 2$ f $x \div 6$ g $1\frac{1}{2} + b$ h $x + y$

2 Write an expression for each situation. Remember to say what any variables represent.

a Joshua is 4 years younger than his sister. What is his age?

b Mike choses a number and multiplies it by 4. What is the product?

c Kira has some sweets. She shares them equally among 3 friends. How many do they each get?

d Tiffany has some money. She gives $5.00 to her sister. How much does she have left?

3 Write an expression for each of the following:

a the sum of a number and 2   b the sum of two different numbers

c 3 less than a number   d 4 more than a number

e three times a number   f twice a number

g half of a number   h a number decreased by 10

i a number multiplied by 3 and then divided by 4

4 Use x, y and z to represent any three numbers. Write an expression for:

a the sum of the three numbers   b the product of x and y

c the sum of x and y d x squared

e 5 more than the sum of x and y f 3 less than the product of x and z

g y less than z h half of the sum of the numbers

5 Write each of these expressions in words.

a $x + 3$ b $y - 2$ c $5 - x$ d $4 \times y$ e $y \div 3$ f $10 \div y$

Looking Back

There are x boys and y girls in a class. Write an expression to represent:
a the total number of students in the class
b the number of boys in the class if 2 new boys enrol
c the difference between the number of boys and girls.

Unit 3 Solve Simple Equations

Let's Think …

What number is represented by *n* in each of these number sentences?

$4 + n = 20$ $4 \times n = 20$ $n \div 4 = 20$ $20 \div n = 2$ $n - 8 = 7$

How did you work out the answers?

The equals sign in an equation is very important. It tells you that both sides of the number sentence have the same value. It also means that you can change the equation as long as you change the parts on either side of the equals sign the same way.

Example *What is the value of x if $14 + x = 20$?*

(You can probably see the answer right away, but follow the working anyway.)

You want to know what *x* is, so you have to get it on its own.

$$14 \;+\; x \;=\; 20$$
$$\downarrow \qquad\qquad \downarrow$$
$$\text{Subtract } 14 \qquad \text{Subtract } 14$$
$$x \;=\; 6$$

To do that, you can subtract 14.

If you subtract 14 from the left-hand side, you need to subtract 14 from the right-hand side as well to keep the two sides equivalent.

You can also use inverse operations to find the missing values.

$$14 + x = 20$$
$$x \rightarrow \boxed{+14} \rightarrow 20$$
$$x \leftarrow \boxed{-14} \leftarrow 20$$
$$\text{so } x = 6$$

Remember: if $2 + 6 = 8$, then $8 - 6 = 2$, so if $x + 6 = 8$, then $8 - 6 = x$

You can check your answers by putting the number you have found in place of the variable and making sure the number sentence is true.

$6 + 14 = 20$, so $x = 6$ is correct.

1 Solve these equations.

a $3 + x = 7$ b $2 + y = 9$ c $2 + y = 10$

d $1 + x = 50$ e $14 - x = 4$ f $5 - x = 1$

g $x + 12 = 23$ h $y + 15 = 50$ i $y + 1\frac{1}{2} = 3$

j $3 \times x = 27$ k $x \times 9 = 54$ l $8 \times y = 24$

m $12 \div x = 4$ n $x \div 4 = 20$ o $35 \div y = 7$

2 Find the value of the variable in each equation.

a $a \times 4 = 48$ b $x + 34 = 82$

c $32 - x = 12$ d $x \div 2 = 20$

e $20 - a = 5$ f $m \times 8 = 40$

g $25 = s + 3$ h $15 \times m = 165$

3 Write a number sentence to represent each problem and solve it to work out the unknown value.

a 5 more than x is 14.

b 10 more than y is 200.

c 3 less than a is 121.

d 45 minus b is 0.

e half of x is 32.

f twice y is 50.

4 Work out the value of x in each of these equations.

a $2 \times x + 11 = 25$ b $x \times 10 - 12 = 48$

c $(x + 10) \times 5 = 500$ d $3 \times x - 15 = 12$

e $5 \times x - 100 = 150$ f $x \times 2 + 14 = 30$

5 Find all the possible values of m and n in these equations.

a $160 \div 20 = m \times n$

b $32 - 17 = m + n$

c $100 - 40 = m \times n$

d $32 + 48 = 2 \times m + n$

Looking Back

Write each equation as a number sentence and solve it.

a 11 plus y is equal to 20.

b 7 less than x is 4.

c The difference between 11 and m is 9. (m is greater than 11.)

d The product of 8 and n is 32.

e Twice m is 4 more than 20.

Topic Review

Talking Mathematics

Make a flow chart to explain how to solve problems in mathematics.

Use some of these phrases.

First …

Then …

Next …

When you have an answer …

Always …

Quick Check

1 What mathematical operation symbol is correct for each word?

   a sum   b product   c quotient   d total
   e more than   f times   g minus   h less

2 Write an expression to represent:

   a the sum of a number and 12
   b the difference between a number and 20, given that 20 is greater
   c the product of a number and 9
   d half of a number

3 Write each of these as a mathematical equation and find the value of n.

   a the sum of n and 4 is 13   b the sum of n and 9 is 17
   c n less 7 results in 12   d the product of n and 8 is 56
   e the sum of twice n and 3 is 21

4 An equilateral triangle has a perimeter of x. Write an expression for the length of a side.

5 Nadia says 'I am thinking of a number. When it is multiplied by 3 and the product is increased by 5, the result is 17'. What is the number she is thinking of?

Topic 16 Length Workbook pages 49–52

Key Words

length
linear
width
kilometre
metre
centimetre
millimetre
decimetre
longer
shorter

▲ Which building is the tallest? Which tree is the shortest? Which units do you use to measure length?

When someone asks 'How tall is it?' or 'How long is it?' they are asking about the **length** of something. In this topic, you are going to estimate and measure lengths using metric units: **millimetres**, **centimetres**, **decimetres**, **metres** and **kilometres**. You will also use decimals and solve problems involving length.

Getting Started

1 Choose the unit that best matches each description: millimetre, centimetre, metre, kilometre.

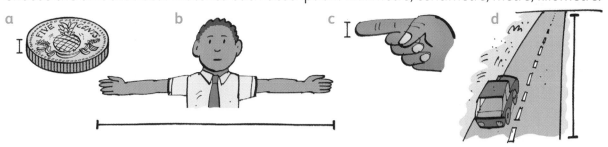

2 Work in groups. In your groups, guess the height of each person. Then use a metre rule or tape measure to measure. What was the difference between the estimated and actual height? Fill in a table like this:

Name	Estimated Height	Actual Height	Difference

Unit 1 Working with Metric Units

Let's Think …

What would be the most appropriate units to measure the following? Choose from mm, cm, m or km.

a The distance from San Andros airport to Orange Cay.

b The distance from the arrivals lounge at the airport to the parking area at the airport.

c The thickness of a passport.

d The height of a stop sign in the airport parking area.

e The length of an airplane.

Length is a linear measure. You use it to measure a line, or the distance between two points. You can compare objects by looking at them to tell which is longer and which is shorter. Guessing the length or width of an object is called estimating. When you use a tape measure or ruler to find the actual length, you are measuring using standard units.

Estimate and Measure

1 Estimate each of the lengths of a chair in cm and then measure the dimensions.

2 Measure the length and width of your mathematics textbook in mm.

3 Look around. What is the longest thing in your classroom? Estimate and then measure its length.

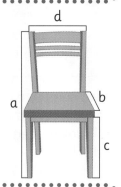

You have already worked with the units of millimetre (mm), centimetre (cm), metre (m) and kilometre (km).

However, there are more units in the metric system:

1 kilometre (km) = 1 000 metres (m)

1 hectometre (hm) = 100 m

1 dekametre (dkm) = 10 m

1 decimetre (dm) = $\frac{1}{10}$ m   'deci-' means 'tenth'

1 centimetre (cm) = $\frac{1}{100}$ m   'centi-' means 'hundredth'

1 millimetre (mm) = $\frac{1}{1000}$ m

The table below shows you the relationships between the metric units of length.

Prefix	Kilo-	Hecto-	Deka-	Metre	Deci-	Centi-	Milli-
	km	**hm**	**dkm**	**m**	**dm**	**cm**	**mm**
In metres (fractions)	1000 m	100 m	10 m	1 m	$1\frac{1}{10}$ m	$\frac{1}{100}$ m	$\frac{1}{1000}$ m
In metres (decimals)	1000 m	100 m	10 m	1 m	0.1 m	0.01 m	0.001 m

Using Decimals

Now you will use decimals to express metric units of length.

$\frac{1}{10} = 0.1$ *one-tenth* $\frac{1}{100} = 0.01$ *one-hundredth* $\frac{1}{1000} = 0.001$ *one-thousandth*

4 Make your own metre stick to show cm, dm and m. You will need:

- Cardboard
- Glue or tape
- A metre ruler
- Pencils and markers

  a Cut a strip of card 1 metre long. If your cardboard is not long enough, stick together several pieces to make a long enough strip.

  b Use the metre ruler to help you measure out the markings on your stick:

  - Make a mark near the one end of your metre stick to show zero.
  - Draw small blue marks to show 1 cm, 2 cm, up to 10 cm.
  - Mark out a red line to show the 10 cm mark. This is also 1 dm.
  - Mark another red line after 2 dm, 3 dm and so on, up to 1 m.
  - Mark the 10 dm point with a black line. This is 1 m.

5 Copy and complete. Use decimals.

  a 1 dm = ☐ m   **b** 2 dm = ☐ m   **c** 5 dm = ☐ m   **d** 1 cm = ☐ m

  e 3 cm = ☐ m   **f** 15 cm = ☐ m   **g** 200 cm = ☐ m   **h** 375 cm = ☐ m

---

Look at this example:

Write 5.82 m as metres, decimetres and centimetres.

5.82 m = 5 m + 0.8 m + 0.02 m

    = 5 m + 8 dm + 2 cm

You call this expanded form, because you have expanded the metres into different units.

---

6 Expand these measurements.

  a 6.89 m = ☐ m + ☐ dm + ☐ cm   **b** 12.43 m = ☐ m + ☐ dm + ☐ cm

  c 1.099 m = ☐ m + ☐ dm + ☐ cm + ☐ mm   **d** 14.032 m = ☐ m + ☐ dm + ☐ cm + ☐ mm

---

Looking Back

1 Write the unit that is equal to:
  a 0.01 m
  b 0.01 km
  c 0.1 cm
  d 0.001 km

2 Write these using decimals.
  a 5 cm 2 mm = ☐ cm
  b 2 m 1 dm 4 cm 7 mm = ☐ km
  c 20 cm 2 mm = ☐ m
  d 65 dm = ☐ m

Unit 2 Creating and Solving Problems

Let's Think …

1 a Which units are bigger than a metre?

   b Which units are smaller than a metre?

2 Your teacher will give you a collection of long, thin objects.

   a Work in your group to arrange the objects
     from longest to shortest.

   b Estimate the length of each object.

   c Measure your objects in cm.

You sometimes need to convert from one unit to another to solve problems.

*A plumber needs 4 pieces of pipe, each one is 8 cm long. She has 4.2 dm of pipe.
How much pipe will she have left after cutting the four pieces?*

$4 \times 8\,cm = 32\,cm$

$4\,dm = \boxed{}\,cm?$

To convert from dm to cm, multiply by 10

$4.2 \times 10 = 42$

$4.2\,dm = 42\,cm$

$42\,cm - 32\,cm = 10\,cm$

She will have 10 cm of pipe left over.

---

1 a Write the units of length in order from smallest to largest.

     metre kilometre decimetre dekametre hectometre millimetre centimetre

   b Now write the abbreviations for each unit, in order from smallest to largest.

2 Write the name of an object that measures about:

   a 1 dm   b 5 mm   c 8 cm   d 0.5 m

3 Solve these problems.

   a A gardener digs a trench that is 62 dm long. How many metres long is the trench?

   b A tree is 85 dm tall. How tall is it in m?

   c An electrician has a roll of 12 m of wire. She must connect five lights to a circuit. Each light requires 45 dm of wire. How much will she have left afterwards?

   d An artist uses strips of wood to make picture frames. He needs 2 strips each with a length of 25 cm, and 2 strips each with a length of 12 cm. He cuts the pieces from one long strip which is 8.8 dm long. What length is left after cutting the pieces?

4 At home, find five pieces of furniture to measure. They may be chairs, tables, desks or any other items.

   ● Estimate each length. Write your estimate.

   ● Measure the item. Write your measurement.

   ● Write the measurement again using a different unit of measurement.

5 A square of chocolate is 8 mm high and 13 mm wide. A slab has 8 rows of squares.

   a How long is the whole slab?

   b What other information would we need to work out how wide the slab is?

13 mm

8 mm

When you compare lengths, you use the <, > and = symbols.

     < means 'less than'

     > means 'greater than'

     = means equal to

Here are some examples:

100 cm = 1 m   125 cm > 1 m   55 mm < 10 cm

6 Copy and complete. Fill in <, > or =.

   a 3 m ☐ 30 cm   b 45 mm ☐ 4.5 cm   c 25 dm ☐ 20 cm

   d 25 m ☐ 625 mm   e 9 cm ☐ 90 mm   f 12 mm ☐ 1 cm

   g 100 m ☐ 1 km   h 9 500 m ☐ 9 km   i 220 cm ☐ 2.3 m

Looking Back

1 A wooden plank is 15 mm thick. I have a stack of 8 planks.

   a How high is the stack?

   b I place the stack on a shelf that is 25 cm high. How many more planks can I fit on the stack before the shelf is full?

2 Copy and complete. Use <, > or =.

   a 18 cm ☐ 180 cm   b 4.5 m ☐ 425 cm   c 33 m ☐ 200 cm

Topic Review

What Did You Learn?

- The basic unit of length is the metre (m). A metre is about the distance from one hand to the other when a child stretches their arms to the sides.
- 1 kilometre (km) is about the length of ten football fields. 1 km = 1 000 m
- A decimetre (dm) is about the width of an adult's hand. 1 dm = 10 cm
- A centimetre (cm) is about the width of your index finger. 100 cm = 1 m
- A millimetre is one thousandth of a metre. 10 mm = 1 cm
- When you convert larger units to smaller units, you multiply.
- When you convert smaller units to larger units, you divide.

Talking Mathematics

1 Look up these words starting with 'deci-' in a dictionary: decimal, decimetre, decibel. Explain why you think they all share the prefix 'deci-'.

2 Sort these words into two groups. Explain how you decided which words go into which group: taller, thicker, shorter, thinner, higher, smaller, narrower, bigger, longer, wider.

Quick Check

1 Write the name of the unit equal to:

a $\frac{1}{10}$ of a metre

b 10 hectometres

c $\frac{1}{10}$ of a dekametre

d $\frac{1}{10}$ of a cm

e $\frac{1}{100}$ of a m

2 Write the following fractions in expanded form using the different units you have learned.

a 7.92 m

b 11.381 m

c 2.713 km

3 A frog jumps 23 cm with each jump.

a How many jumps would it take the frog to cover more than 2 m?

b After the last jump, what was the total distance the frog covered?

4 Alex records how far her toy car goes with each push.

Distance Covered			
31.5 cm	27.9 cm	42.4 cm	39.8 cm

a Work out the total distance covered by the car.

b Express your answer in metres.

c How much further must the car go in order to cover 3 m?

Topic 17 Multiplying and Dividing

Workbook pages 53–55

Key Words

multiply

product

divide

quotient

divisibility

inverse

remainder

▲ Mrs Carlson hand weaves these patterned strips to make baskets. She uses 7 metres to make a large basket. How many metres would she need to weave if she got an order for 25 baskets? If she wove 135 metres of patterned strip, how many baskets could she make?

Earlier this year, you revised multiplication and division facts and you made fact families using **inverse** operations. You also learned about factors and multiples. Now, you are going to use what you know to **multiply** and **divide** larger numbers. There are some rules that you can apply to see whether a number can be divided exactly by another and you are going to learn to use these to make division easier. You will also learn some quick methods for multiplying and dividing by 10, 100 and 1 000 as well as other multiples of 10.

Getting Started

Talk about these two problems in groups.

● How do you know when a problem involves multiplication?

● How do you know when to divide to solve a problem?

● What number sentences could you write to solve each of these problems?

A car uses 8 litres of gas for every 115 km it travels. How many litres of gas would it use to travel 725 km?

Johnny loves computers and he can type very fast. He estimates that he can type 50 words per minute. How many words is this per half hour?

Unit 1 Revisit Multiplying and Dividing

Let's Think …

- Explain how knowing that $6 \times 5 = 30$ can help you work out that $300 \div 6 = 50$?
- Given that $228 \div 12 = 19$, what is 12×19?
- How do you know that without doing any calculation?

Remember that multiplication and division are inverse operations.
You can use multiplication facts to work out division facts.
$12 \times 5 = 60$ So, $60 \div 5 = 12$ and $60 \div 12 = 5$
You can use division facts to work out multiplication facts.
$138 \div 3 = 46$ So, $3 \times 46 = 138$

---

1 Write the answers to these facts as quickly as you can.

a 7×4	b 7×6	c 9×8	d 2×4
e 5×6	f $63 \div 7$	g $40 \div 5$	h $18 \div 3$
i $49 \div 7$	j $63 \div 9$	k 5×9	l 10×10
m 7×9	n 4×4	o 3×6	p $48 \div 4$
q $12 \div 6$	r $20 \div 10$	s $32 \div 8$	t $15 \div 3$
u 9×9	v $36 \div 6$	w $81 \div 9$	x $56 \div 8$

2 These divisions all leave a remainder. How quickly can you write the answers?

a $22 \div 9$	b $34 \div 7$	c $46 \div 6$
d $58 \div 8$	e $79 \div 6$	f $68 \div 9$
g $81 \div 7$	h $67 \div 9$	i $60 \div 7$
j $65 \div 8$	k $47 \div 7$	l $44 \div 6$

3 Choose three factors from the box and find their product. Can you make 9 different products?

2	3	7	8	10

4 Answer these questions.

a How many sixes are in 63?

b How can you share 56 evenly among 7 people?

c What is $\frac{1}{3}$ of 27?

d What is 120 divided by 10?

5 For each fact that is given below, write a fact family with the four related facts.
  a $7 \times 26 = 182$ b $5 \times 39 = 195$
  c $234 \div 9 = 26$ d $950 \div 19 = 50$

6 Malia has made some fact family cards. One fact is missing from each family. Find the families and write the missing facts.

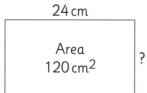 $120 \div 4 = 30$

$41 \times 7 = 287$

$588 \div 6 = 98$

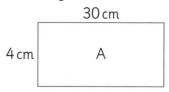 $4 \times 147 = 588$

$147 \times 4 = 588$

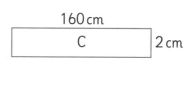 $6 \times 98 = 588$

$98 \times 6 = 588$

$30 \times 4 = 120$

$7 \times 41 = 287$

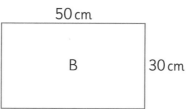 $287 \div 7 = 41$

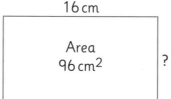 $588 \div 4 = 147$

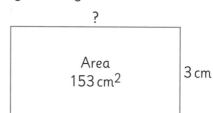 $120 \div 30 = 4$

7 Find a fact above to solve each of these problems. Write the answers only.
  a There are 41 rows of 7 seats. How many seats is this in total?
  b How many groups of 4 can you make from 120?
  c An air ticket costs $98.00. If the total cost of tickets is $588.00, how many tickets were purchased?
  d How many air tickets costing $147.00 each can you buy with $588.00?

8 The area of a rectangle is its length multiplied by its width. What is the area of each of these rectangles?

30 cm
4 cm   A

50 cm
B   30 cm

160 cm
C   2 cm

9 The area of these rectangles is given. Work out the missing side length.

24 cm
Area 120 cm²   ?

16 cm
Area 96 cm²   ?

?
Area 153 cm²   3 cm

Looking Back

$6 \times 8 = a$ $6 \times b = 48$ $c \times 8 = 48$

  a What are the values of a, b and c?
  b Which operation did you use to find b and c? Why?
  c If $x \div 6 = 9$, what would you do to find the value of x? Why?

Unit 2 Divisibility Rules

Let's Think …

- How can you tell whether a number can be exactly divided by 2?
- Can these numbers be divided by 2 with no remainder?

  124   865   988   10 000   954   179   14 309

When a number can be divided by 2 with no remainder, you say it is divisible by 2.

Here are some simple rules that you can use to decide whether a number is exactly divisible by another number.

A number is divisible by …

2 if it is even (ends in 0, 2, 4, 6 or 8)	**3** if the sum of its digits is a multiple of 3	**4** if the last two digits are a multiple of 4
5 if the last digit is 0 or 5	**6** if it is even *and* divisible by 3	**8** if the last two digits are a multiple of 8
9 if the sum of its digits is a multiple of 9	**10** if it ends in 0	

You will notice that there is no rule for 7. There is no easy way to work this out by looking at the digits.

Read through the examples to see how these rules work.

Example 1 Does 4 divide exactly into 2 578?

Rule: Last two digits must be a multiple of 4.

Check by dividing:

4 is not a factor of 78, so 78 is not a multiple of 4.

2 578 cannot be divided exactly by 4.

1 Which of these numbers are exactly divisible by 3?
   a 12   b 148
   c 3 543   d 10 230

2 Test these numbers to see if they are divisible by 6. Give a reason if the number is not divisible by 6.
   a 128   b 129
   c 234   d 1 242
   e 906   f 315
   g 912   h 2 391

3 Without doing each division, say whether there will be a remainder or not.
   a $12 430 \div 5$ b $9 246 \div 4$ c $9 236 \div 4$ d $12 375 \div 3$
   e $49 230 \div 9$ f $3 608 \div 8$ g $12 540 \div 10$ h $19 995 \div 10$
   i $23 460 \div 6$ j $3 457 \div 2$ k $3 006 \div 3$ l $23 458 \div 5$
   m $7 341 \div 6$ n $52 467 \div 3$ o $1 345 \div 9$ p $23 128 \div 8$

4 What is the smallest whole number that is divisible by:
   a 2, 4 and 5
   b 3, 5, 8 and 9

5 Can $99.40 be shared equally by 4 people? Give a reason for your answer.

6 Three classes collect money for charity. They collect $42.00, $35.50 and $62.50. Can they make the total amount using a combination of five- and ten-dollar bills?

Looking Back

Use the digits 0, 1, 3, 5 and 8 to make:
a two numbers that are divisible by 5.
b four different numbers that are divisible by 10.
c two four-digit numbers that are divisible by 4
d four five-digit numbers that are divisible by 2.
Can you make a number that is divisible by 50?

Unit 3 Multiple and Powers of Ten

> **Let's Think …**
>
> Look at these calculations.
> What do you notice?
>
> $23 \times 10 = 230$ $23 \times 100 = 2\,300$ $23 \times 1\,000 = 23\,000$
>
> $14\,000 \div 10 = 1\,400$ $14\,000 \div 100 = 140$ $14\,000 \div 1\,000 = 14$

You know that each place on the place value table is 10 times greater than the value to its right. You can use this fact to find quick methods of multiplying and dividing by powers of ten.

Remember $10 \times 1 = 10$ $10 \times 10 = 100$ $10 \times 10 \times 10 = 1\,000$

          1 place   2 places   3 places

To multiply by 10, 100 or 1 000 you move the digits 1, 2 or 3 places to the left and write zeros in the empty places.

Example 1   Multiply 19×10 19×100 $19 \times 1\,000$

T Th	Th	H	T	O
			1	9
		1	9	0
	1	9	0	0
1	9	0	0	0

19×10 Move digits 1 place left.
Write 0 in the ones place as placeholder.

19×100 Move digits 2 places left.
Write 0 in the ones and hundreds places as placeholder.

$19 \times 1\,000$ Move digits 3 places left.
Write 0 in hundreds, tens and ones places as placeholder.

Example 2   Divide $1\,200 \div 10$ $1\,200 \div 100$ $1\,200 \div 1\,000$

Th	H	T	O	.	T $\left(\frac{1}{10}\right)$
1	2	0	0		
	1	2	0		
		1	2		
			1	.	2

$1200 \div 10$ Move digits 1 place right.
Ignore zeros that 'fall off' the whole number table.

$1200 \div 100$ Move digits 2 places right.
Ignore zeros that 'fall off' the whole number table.

$1200 \div 1\,000$ Move digits 3 places right.
Your answer is a decimal. You cannot ignore 0.2!

These methods are also useful for multiplying and dividing by multiples of 10 using factors.

Example 3

300×6 Think of this as $3 \times 100 \times 6$

$3 \times 6 = 18$ $18 \times 100 = 1\,800$

Example 4

$3\,600 \div 30$ You know that $30 = 10 \times 3$ so you can divide by 10 then by 3

$3\,600 \div 10 = 360$ $360 \div 3 = 120$

1 Try to do these calculations mentally. Write the answers only.

   a 5×10 b 23×10 c 87×10 d 124×10 e 200×10

   f 654×10 g $1\,235 \times 10$ h $7\,500 \times 10$ i $8\,999 \times 10$

2 Determine these products mentally. Check your answers with a calculator.

   a 8×100 b 23×100 c 80×100 d 325×100 e 650×100

   f 854×100 g $1\,230 \times 100$ h $2\,300 \times 100$ i $7\,045 \times 100$ j $23 \times 1\,000$

   k $29 \times 1\,000$ l $36 \times 1\,000$ m $124 \times 1\,000$ n $328 \times 1\,000$ o $450 \times 1\,000$

3 Divide mentally. Write the answers only.

   a $540 \div 10$ b $700 \div 10$ c $350 \div 10$ d $1\,200 \div 10$ e $7\,650 \div 10$

   f $3\,000 \div 10$ g $1\,200 \div 100$ h $4\,500 \div 100$ i $9\,900 \div 100$ j $12 \div 10$

4 Use factors to find these products.

   a 3×20 b 6×40 c 8×80 d 9×50 e 4×80

   f 6×90 g 200×9 h 5×300 i 400×8 j 900×3

5 Are these statements TRUE or FALSE?

   a $34\,000 \div 1\,000 = 3\,400 \div 100$ b $40\,000 \div 100 = 400\,000 \div 100$

   c $9\,200 \div 100 = 19\,200 \div 10$.

6 Can you work out these products mentally using the same rules?

   a 1.2×10 b 3.5×100 c $2.8 \div 10$

Looking Back

Calculate mentally.

a $279 \times 1\,000$ b 1.4×100 c 208×10 d $2\,780 \div 10$

e $2\,780 \div 100$ f 9.9×100 g 17×20 h 9×800

Unit 4 Multiply Larger Numbers

Let's Think …

A class is given this problem.

Sondra has 19 boxes of crayons. Each box contains 12 crayons. How many does she have in all?

Look at these three solutions.

Which one would you have used? Why?

Which one seems easiest? Why?

Solution A

$19 \times (10 + 2)$

$19 \times 10 = 190$

$19 \times 2 \ = \ \underline{\ \ 38}$

$\qquad\qquad \underline{228}$

Solution B

$\begin{array}{r} 19 \\ \times\ 12 \end{array}$

18	2×9
20	2×10
90	10×9
$\underline{100}$	10×10
$\underline{228}$	

Solution C

$\times$	10	9
10	100	90
2	20	18

$\quad\ 120$

$\quad\ \underline{108}$

$\quad\ \underline{228}$

You learned how to multiply larger numbers last year using the grid method and the long multiplication method. Read through these examples to revise these two methods of working.

Calculate 23×148

Grid method

Estimate: $140 \times 20 = 140 \times 10 \times 2 = 1\,400 \times 2 = 2\,800$

$\times$	20	3		
100	2 000	300	$\rightarrow$	$2\overset{1\ 1}{3}00$
40	800	120	$\rightarrow$	920
8	160	24	$\rightarrow$	184
				$\underline{3\,404}$

Write the number in expanded notation on the grid.

Multiply to fill each cell.

Add the totals in each row to get the product.

Calculate 23×148

Long multiplication method

Estimate: $140 \times 20 = 140 \times 10 \times 2 = 1\,400 \times 2 = 2\,800$

```
    148
  ×  23
  _____
  ₁ 24      (3 × 8)
   120      (3 × 40)    Then multiply each digit in the top number by 20.
   300      (3 × 100)
   160      (20 × 8)
 ₁ 800      (20 × 40)   Add to find the product.
 2 000      (20 × 100)
  _____
 3 404
```

Multiply each digit in the top number by 3, remember the place values.

1 Estimate and then calculate the products. Use the method that you find most efficient.

   a 234×65 b 378×14 c 534×75 d 903×66

   e 220×48 f 297×70 g 98×322 h 12×350

2 There are 24 bottles in a crate. How many bottles will there be in 368 crates?

3 A family saves $245.00 per month. How much will they save in a year?

4 Bananas are transported from the farm to the packing plant by truck. Each truck carries 576 hands of bananas.

   a The mean mass of a hand of bananas is 480 grams. What is the mean mass carried by each truck?

   b How many bananas are carried by 14 trucks if they each make two trips?

5 A factory makes 321 T-shirts per week. How many will they make in a year if they close down for 3 weeks each summer?

6 Kayla calculates that she takes 960 steps to get to school and back each day. How many steps will she take to school and back in four weeks?

7 A school principal lives 11 km from school. During a year, she travels to school and back about 275 times. Work out how many kilometres this is in total.

Looking Back

Find the product of:

a 41 and 24   b 65 and 14   c 158 and 18   d 124 and 34

Unit 5 Dividing Larger Amounts

Last year, you learned how to divide by a 1-digit divisor using a division house symbol and carrying numbers.

This method is called short division. Here are two examples.

$$\begin{array}{r} 14 \\ 9\overline{)\ 12\,^36} \end{array} \qquad \begin{array}{r} 17\ r^5 \\ 7\overline{)\ 12^59} \end{array} \qquad 7 \times 8 = 56$$

When you divide by a 2-digit number, it is easier to keep track of your working if you use a longer method of writing out your calculation. Look at these two examples to see how to do this.

Example 1   249 ÷ 21

Step 1

$$\begin{array}{r} 1 \\ 21\overline{)249} \\ -\,21 \\ \hline 3 \end{array}$$

Divide the tens
21 × 1 = 21
Subtract 24 − 21 = 3
21 cannot go into 3

Step 2

$$\begin{array}{r} 1\,1\ r\,18 \\ 21\overline{)249} \\ -\,21\!\downarrow \\ \hline 39 \\ 21 \\ \hline 18 \end{array}$$

Bring down the ones
Divide
21 × 2 = 42 ✗ too high
21 × 1 = 21
Subtract 39 − 21 = 18
18 < 21 so it is the remainder

Example 2   1 376 ÷ 12

$$\begin{array}{r} 1\ 1\ 4\ r\,8 \\ 12\overline{)1376} \\ -12\!\downarrow \\ \hline 17 \\ -12\!\downarrow \\ \hline 56 \\ -48 \\ \hline 8 \end{array}$$

Bring down tens
Bring down ones
12 × 4 = 48
Subtract 56 − 48 = 8
8 < 12 so it is the remainder

This method is called long division.

1 Calculate. Show your working out.
  a 868 ÷ 15   b 636 ÷ 21
  c 906 ÷ 52   d 456 ÷ 16
  e 987 ÷ 41   f 843 ÷ 27

2 Use long division to find the following quotients.
  a 3412 ÷ 15   b 6712 ÷ 31
  c 9873 ÷ 18   d 1235 ÷ 24
  e 2346 ÷ 21   f 1987 ÷ 23

3 The mass of 37 sacks of cement is 3 885 kg. What is the mass of one bag?

4 What is the quotient and remainder if the product of 168 and 46 is divided by 29?

Looking Back

Karen is running a virus check on her computer. The display shows this information:

> ## Progress
>
> Status: Scan progressing...
> Product: Fraud.Sysguard
> Estimated time left: 14365 minutes...

a How many hours is this going to take to finish?
   She looks again later and the display shows this information.

> ## Progress
>
> Status: Scan progressing...
> Product: Fraud.Sysguard
> Estimated time left: 5414 minutes...

b How much time passed (in hours) from the first display to the second?

Topic Review

What Did You Learn?

- Multiplication and division are inverse operations. You can use one fact to work out a fact family.
- You can use place value to multiply and divide by powers and multiples of 10 mentally.
- You can use divisibility rules to decide whether a number is exactly divisible by another without doing the division.
- To multiply larger numbers you can use place value, a grid or a column method.
- Always estimate by rounding before you calculate.

- The answer to a division is called the quotient. The number being divided is called the dividend and the number you are dividing by is called the divisor.
- When one number does not divide exactly into another, you are left with a remainder.
- You can use short division as a written method to divide two- and three-digit numbers by a single digit.
- When you are dividing by a two-digit number, it is more efficient to record your work using long division.

Talking Mathematics

Micah asks, 'What number divided by 63 will give a quotient of 48 and a remainder of 16?'

Jessica says, 'You can work out the answer by multiplication'. What do you think Jessica means?

Is she completely correct?

What is the missing number?

Quick Check

1 Tony has 2 462 elastic bands. He packs these into boxes each containing 80 elastic bands. How many boxes can he fill? How many elastic bands will be left over?

2 Sharyn has 320 points. Andy has 100 times as many. How many does Andy have?

3 A rectangle is ten times as long as it is wide. If it 23 cm wide, calculate its area.

4 The ferry from Nassau to Paradise Island costs $2.00. If there are 35 passengers per trip and the ferry crosses 9 times, how much money will they have collected in ticket fees?

5 Mrs Norris can pack mangoes into bags containing 12, 14, 15, 16, 18 or 20. Which size bag is best for 1 265 mangoes seo that the fewest possible mangoes will be left over?

Topic 18 Order of Operations Workbook pages 56–57

Key Words
operation
grouping symbols
brackets
multiply
divide
add
subtract

▲ Think about some of the rules that are important in our lives; for example, in The Bahamas, cars drive on the left-hand side of the road. Why do you think we have rules like that? What happens if people do not obey the rules? Can you think of any situations where drivers might drive on the right in The Bahamas?

In real life we follow rules, but we also do things in an agreed order; for example, most people agree that it is sensible to toast the bread before you spread butter and jam on it, and it is agreed all over the world that you look both ways before you cross a busy road. In mathematics, we also need to do things in an agreed order. Mathematicians like order, and they have agreed that when there is more than one **operation** in the same number sentence, you have to do them in a particular order to avoid confusion and getting different answers. You are going to learn the rules for ordering operations and then you are going to practice using them.

Getting Started

Look at the picture.

- Why did the girls get different answers?
- If the original problem was: Josh has 3 five dollar bills and 2 dollar coins. How much money does he have? Which is the correct answer? Explain your thinking.
- Can you think of a problem that would make the other answer correct?

Unit 1 Grouping Symbols

Let's Think …

Read the problems.

Jayden has 2 loose sweets and five packets with 6 sweets each. How many does he have in all? 	Jackson puts 2 red and 5 green sweets into packets. How many sweets does he need to fill 6 packets? 

How are they different?

Write a number sentence for each and solve it.

What do you notice?

In mathematics, different types of brackets are used to group operations. Each type has a name, but they are often just called brackets. You do the operations inside sets of brackets first.

Parentheses ( ): $2 \times (3 + 4)$ This means you add $3 + 4$ and then multiply the result by 2.

$$2 \times (3 + 4)$$
$$= 2 \times 7$$
$$= 14$$

There are also other grouping symbols, but you only need to work with brackets in this topic.

The key rule of order of operations is that you always do operations inside grouping symbols first.

1 Do these calculations correctly. Show your working.

  a $(3 + 3) \times 10$
  b $(18 - 3) \div 5$
  c $25 - (5 + 7)$

  d $8 \times (4 + 2)$
  e $(3 + 4) \times 7$
  f $(20 - 12) \div 8$

  g $12 + (42 \div 7)$
  h $(20 - 4) \div 4$
  i $7 \times (11 - 6)$

  j $(3 - 2) \times 4$
  k $(6 + 7) \times 3$
  l $(12 - 8) \times 9$

2 Calculate correctly.

a $(3+3) \times (14-4)$ b $(9-5) \times (2+7)$ c $(4+16) \div (12-7)$

d $(26+4)-(3 \times 3)$ e $(10 \times 10) \div (5+5)$ f $(3-2) \times (8+4)$

g $(4 \times 6)+(18 \div 6)$ h $(9-5) \times (19-10)$ i $(17-8) \div (3 \times 3)$

j $(4+8) \times (17-16)$ k $(14-13)+(20 \div 20)$ l $(7+8) \times (12 \div 12)$

3 Work from the innermost to the outermost sets of brackets to correctly calculate these answers.

a $3+[8-(2 \times 4)]$ b $15-[4+(6 \div 2)]$ c $[(24 \div 6)+5]-7$

d $48-[25-(4 \times 6)]$ e $[8 \times (5+2)] \times 10$ f $[(14 \div 2)+2] \times 3$

g $16 \div [54-(5 \times 10)]$ h $[100 \div (5+5)] \times 2$ i $2 \times [35 \div (4+3)]$

j $[(7+5) \times 3] \div 9$

4 Write a number sentence with parentheses to correctly represent each problem. Solve it and write the answers.

a I have 6 packets with 5 markers in each and 3 loose markers. How many in all?

b Subtract the quotient of 21 divided by 3 from 19. How much is left?

c Take 2 from 8 and then multiply the result by 5. What is the answer?

d Tickets for a show cost $12.00. If you buy four tickets, you get $3.00 off each price. How much will it cost for 4 tickets?

Looking Back

Calculate the answers.

a $4 \times (13-5)$ b $5 \times (6-3)$

c $(6 \times 8)-(10+10)$ d $(6+7)-(12 \div 4)$

e $[8 \div (16-14)] \times 3$ f $[11 \times (30 \div 10)]+5$

Unit 2 Order of Operations

Calculations like the one above can be done in different ways.

Some people will work from left to right and *subtract* first, then *divide*.

$10 - 6 \div 2$

$= 4 \div 2$

$= 2$

Others will do the division first and then subtract the result.

$10 - 6 \div 2$

$= 10 - 3$

$= 7$

Each way gives a different answer and this can be very confusing, so we need to apply some rules to make sure that everyone gets the same result.

The rules for working are very simple. You already know the first rule!

- Do operations inside brackets first.

- Next multiply or divide in the order in which they appear (work from left to right).

- Then add or subtract in the order in which they appear (again, work from left to right).

When you apply the rules, there is only one correct answer because there is only one correct order of working.

Example 1

$4 \times (4 - 2) + 1$	Parentheses first: $4 - 2 = 2$
$= 4 \times 2 + 1$	Multiply next: $4 \times 2 = 8$
$= 8 + 1$	Then add: $8 + 1 = 9$
$= 9$	

Example 2

$2 + 18 \div 3 \times 7$	Divide before you add: $18 \div 3 = 6$
$= 2 + 6 \times 7$	Multiply before you add: $6 \times 7 = 42$
$= 2 + 42$	Add: $2 + 42 = 44$
$= 44$	

1 Read the statements. Say whether they are TRUE or FALSE.

   a For $2 + 4 \times 5$, we would do $2 + 4$ first.

   b For $(2 + 4) \times 5$, we would do $2 + 4$ first.

   c For $10 - 4 \times 2$, we would do 4×2 first.

   d For $6 + 12 \div 3$, we would divide before we add.

   e For $5 \times 10 \div 2 + 3$, we would do $10 \div 2$ first.

   f For $20 - 6 \times 3 + 15$, we would subtract first.

2 Rewrite each calculation from question 1 and solve it correctly.

3 Apply the rules for order of operations to find the answers to each calculation.

   a $(8 - 2) + 4$ b $18 - 4 \times 2 - 3$ c $(12 - 9) \times (24 - 22)$

   d $14 - 21 \div 3$ e $3 + 2 \times 8$ f $29 - 2 \times 10$

   g $3 \times 4 - 2$ h $(20 + 5) \times 3$ i $5 \times 4 + 30 \div 10$

   j $24 \div 8 \times 6 - 5$ k $25 + 14 \div 2 - 20$ l $3 \times 3 - 4 \times 2$

   m $15 \div 3 - 3 - 2$ n $7 - 24 \div 6$ o $8 \times 3 \div 4$

   p $5 + 36 \div 6$ q $54 - 3 \times 8$ r $40 - 10 \times 3$

4 Work carefully and follow the rules to do these calculations.

   a $4 + 8 \times 2 - 25 \div 5 \times 2 + 4$ b $100 - 10 \times 5 - 30 \div 3 \div 5 + 12$

   c $3 + 21 \div 7 + 4$ d $3 + 18 \div 9 + 7$

   e $18 + 5 \times 1 - 6 \times 3$ f $12 - 2 - 5 \times 2 + 8$

   g $8 - (13 - 8 - 2) \times 3$ h $(6 + 4 \times 5) - 2 \times 4$

5 You can use exactly four 4s together with any operation signs and parentheses if you need them to calculate all the numbers from 0 to 9.

Here are two ways of calculating 0.

$4 + 4 - 4 - 4 = 0$ or $(4 \times 4) - (4 \times 4) = 0$

Work with a partner to find at least one way of calculating all the numbers from 1 to 9.

Looking Back

Miss Rolle gave her class this calculation:

$$2 \times (8 - 3) - 2 \times 2$$

Kamaya got 22   Toniqua got 6

Sam got 16   Leroy got 12

Who was correct?

Try to work out what the others did wrong.

Topic Review

What Did You Learn?

- In mathematics, you have to know and follow the rules to calculate correctly.
- Grouping symbols show you what operations to do first.
- Parentheses ( ), brackets [ ] and braces { } are all grouping symbols.
- When there are brackets inside brackets you work from the inside outwards.
- Operations are done in the following order: Brackets first $\rightarrow \times$ and $\div$ next (from left to right) $\rightarrow +$ and $-$ last (from left to right).

Talking Mathematics

Describe in words the steps you would take to solve each of these calculations.

a $20 \times 3 + 4$

b $4 + 8 \div 2$

c $5 + 6 \times 2 - 2$

d $15 - 3 \times (12 - 6)$

Quick Check

1 Check Martin's homework answers and correct any that he got wrong.

a. $3 + 2 \times 8 = 19$

b. $39 - 2 \times 6 = 222$

c. $3 \times 4 - 2 = 10$

d. $(20 + 5) \times 3 = 75$

e. $(3 \times 2) + 7 \times 3 = 39$

f. $5 \times 4 + 30 \div 10 - 3 = 2$

2 Calculate.

a $(1 + 4) \times 20 \div 5$ b $1 + (4 \times 20) \div 5$

c $6 \times (4 \div 2) \times 3$ d $(6 \times 4) \div 2 \times 3$

e $3 + (5 - 2) \times 3$ f $3 + 5 - (2 \times 3)$

g $50 + 10 \div 10$ h $(50 + 10) \div 10$

i $10 \times 5 + 0$ j $10 \times (5 + 0)$

k $42 \div 6 \times (3 - 3)$ l $42 \div 6 \times 3 - 3$

3 Use the four digits of the current year together with the operation signs and brackets if you need them to make up 10 different calculations. Each calculation must have at least two operations. Exchange with another student and do each other's calculations. Check each other's answers when you have finished.

Topic 19 Working with Time Workbook pages 58–62

▲ This is the clock on the St Matthew's Anglican Church in Nassau.
Where have you seen clocks that look like this?

In the past, people would check the time by looking at clocks like the one in the picture. Can you think of the different devices people use to read the time today? In this topic, you will tell the time on analogue, digital and 24-hour clocks. You will estimate, compare and measure the time needed to complete tasks. You will convert between units of time, and solve problems involving time.

Key Words

occurrence
duration
minute
second
year
day
century
decade
Système International (SI)
elapsed

Getting Started

1 Think of something you can do in one second.
2 About how long is one of your lessons at school?
3 Which is longer: your first or second break at school?
4 What takes longer: putting on your shoes or eating breakfast?
5 If you need to know what time it is at home, how do you usually find out?
6 What kind of clocks have you seen before?

Unit 1 Estimate, Compare and Measure Time

You can talk about time in different ways. You can describe when something happened, or when it will happen.

- *My birthday is in August.*

- *School starts at 9:00 a.m.*

- *My friends will arrive tomorrow.*

This tells you the time of occurrence, or when it happens.

You can also ask 'How long does it take?' The answer to this question tells you the duration, or how long something takes from start to finish.

- *The movie starts at 3:00 p.m. and ends at 5:15 p.m.*

- *The carnival runs from Friday afternoon until Sunday evening.*

- *I will stay with my aunt from July until September.*

1 minute (min) = 60 seconds (s)	1 leap year = 366 days
1 hour = 60 min	1 year = 12 months or about 52 weeks
1 day = 24 hours	1 decade = 10 years
1 week = 7 days	1 century = 100 years
1 year = 365 days	

To convert from a smaller unit to a larger unit, use division.

180 seconds = ☐ minutes

180 ÷ 60   = 3

180 seconds = 3 minutes

To convert from a bigger unit to a smaller unit, use multiplication.

5 hours = ☐ minutes

5 × 60 = 300

5 hours = 300 minutes

1 How long do these activities usually take? Write your answer in hours or minutes.

  a Watching TV.

  b Drinking a glass of water.

  c Putting on your clothes in the morning.

  d Playing on a computer, phone or tablet.

  e Playing a ball game such as cricket or netball.

  f Drying off after swimming.

2 Choose the unit that best matches how long it would take.

  a One (minute/year/century) ago, no one had mobile phones.

  b After each birthday, I have to wait a (decade/year/week) until the next one.

  c Max is 10 years old today. It will be a (leap year/century/decade) before he is 20 years old.

  d It takes one (day/year/decade) for the Earth to travel around the Sun.

  e It takes one (day/year/decade) for the Earth to turn on its axis.

3 Convert between these times.

  a 3 hrs = ☐ min   b 4 d = ☐ hrs   c 8 hrs = ☐ min

  d 2.5 d = ☐ hrs   e 185 seconds = ☐ min   f 32 hrs = ☐ d

In different places, people write the date differently. Each system is known as a format. The British date format is day-month-year (smallest unit to largest unit).

British format: 03 January 2015 or 03/01/2015.

The American date format is month-day-year.

American format: January 03, 2015 or 01/03/2015.

Système International (SI) is an international format, which follows the pattern year-month-day:

SI format: 2015/01/03

---

4 a Write the birthdays of each person in your family using SI format.

  b Which of the formats looks most sensible to you? Why?

5 Write the following dates in SI format.

  a one day after 2017/05/03   b one year before 2008/09/08

  c one decade after 1989/12/01   d one century before 1965/02/11

Looking Back

1 Describe the duration of:

  a one lesson at school   b one term of school   c the time from sunrise to sunset.

2 Write today's date in each of the following formats: British, American and SI units.

3 What unit describes:

  a 10 decades   b 12 months   c 24 hours

Unit 2 Telling the Time

Let's Think ...

- How long does it take for the minute hand to do a full turn on the clock?
- How long does it take for the hour hand to do a full turn?
- How long does it take for the second hand to do a full turn?
- At half past 2, where will the hour hand point? Why?

You can say the time in different ways. For the analogue clock above, you can say:

  2 hours, 25 minutes and 40 seconds

or   25 minutes and 40 seconds after 2

or   20 seconds before 2:26.

You can also use a digital clock to tell the time.
A digital clock shows time using only numbers.

11:30 p.m. means half past 11 in the evening.
11:30 a.m. means half past 11 in the morning.

Hour  a.m. or p.m. tell
you if the time is
in the morning or
Minutes after hour   afternoon/evening

This clock uses the 24-hour clock. Instead of
switching from a.m. to p.m. after 12 o'clock, the hours continue to
13:00 (1 o'clock), 14:00 (2 o'clock), and so on up to 23:00 (11 o'clock).

On the 24-hour clock, each day starts at 00:00, which is midnight. The clock in the
picture shows 23:30, which is the same as 11:30 p.m.

$$23:30$$

---

1 Write each analogue time in two different ways.

2 Write these times using digital notation.

   a 23 minutes past 5 in the afternoon.   b Half past one in the morning.

   c Five minutes before six o'clock in the morning.   d Noon.

3 Write these times in words.

   a 12:45 p.m.   b 2:55 a.m.   c 1:15 p.m.   d 9:28 p.m.

4 Write the times from question 3 in 24-hour notation.

Solving Problems

Elapsed time is the time that passes from the beginning of an event until it finishes. Look at this example.

Mila starts her piano lesson at 3:30 p.m. The lesson lasts 45 minutes. What time does it finish?

	Hours	Minutes
	3	30
+		45
=	3	75

60 minutes = 1 hour

75 − 60 = 15

=   4 hours 15 minutes

She finishes at 4:15 p.m.

The **start time** is 3:30 p.m. The **elapsed time** is 35 minutes. The **end time** is 4:15 p.m.

Can you think of other ways to work out the answer?

How could you count forward on a clock?

5 Work out the end time for each activity.

  a I go for a run at 11:25 a.m. I run for an hour and ten minutes.

  b A TV show starts at 8:05 p.m. It lasts for 45 minutes.

  c Sports day starts at 8:15 a.m. It finishes 5 and a half hours later.

6 Work out the start times.

  a I get home at 5:10 p.m. after a 55 minute car journey. What time did I start the journey?

  b A movie finishes at 7:55 p.m. It was 2 hours 35 minutes long. What time did it start?

7 How long did it take? Work out the elapsed times.

  a James put a cake in the oven at 11:07 a.m. and took it out at 11:42 a.m. How long did it bake for?

  b Rebecca started a car journey at 10:32 a.m. and arrived at her destination at 12:05 p.m. How long was the journey? (Remember, after 12 noon, we change from a.m. to p.m., so 12:05 p.m. is daytime!)

  c Jenna started her run at 6:20 p.m. and finished at 7:07 p.m. How long did she run for?

Looking Back

Write your own rule for calculating:

● start time of an event

● end time of an event

● elapsed time.

Topic Review

What Did You Learn?

- The relationships between units of time: seconds, minutes, hours, days, weeks, years, decades and centuries.
- The time of occurrence describes the time when something happens; for example, today or at 14:30 or in October.
- The duration is how long something takes; for example, two hours or 42 seconds or a decade.
- There are different formats for recording dates, including British, American and SI formats. The SI format is set out as year-month-day: 2017/11/29.
- To convert from a smaller unit of time to a larger unit, use division.
- To convert from a larger unit of time to a smaller unit, use multiplication.
- Elapsed time is the time that passes between the start time and end time of an event.

Talking Mathematics

What do these expressions about time mean? Write your own explanations. You can use a dictionary or the internet to help you.

> There is no time like the present.

> I am just killing time.

> We should call it a day.

> It is crunch time.

> This can help you to save time.

Quick Check

1 Write the matching pairs of times.

> 8 o'clock at night   ten to 1   9:00 p.m.
> 18:00   6:00 p.m.   half past 12
> 20:00   00:30   12:50   21:00

2 How many days in:
   a 2 weeks?   b a leap year?

3 My grandfather has lived for 5 decades and nine years. How old is he?

4 Copy and complete.
   a 6.5 minutes = ☐ seconds
   b 3.5 days = ☐ hours
   c 72 hours = ☐ days

5 Write today's date in SI format.

6 Ali starts her homework at 3:25 p.m. and finishes at 4:50 p.m. How long did she spend on it?

Topic 20 Calculating with Fractions and Decimals

Workbook pages 63–64

Key Words

fraction

mixed number

equivalent

regroup

decimal

decimal point

▲ $1\frac{1}{2}$ chicken pies and $1\frac{3}{4}$ mutton pies are left after a school function. The principal asks how many pies in total are left. What would you answer? How did you work that out?

You already know how to add and subtract **fractions** and **mixed numbers** and you have solved problems involving **decimal** amounts of money. In this topic, you are going to learn how to **regroup** fractional amounts and how to use place value to add and subtract decimals. You will apply what you learn to solve problems involving fractions, mixed numbers, decimals and money amounts.

Getting Started

1 Look at the diagrams carefully.

   a How many quarters are there?

   b Add the quarters. Write the answer as a mixed number.

   c How many fifths are there?

   d How could you write the total number of fifths as a mixed number?

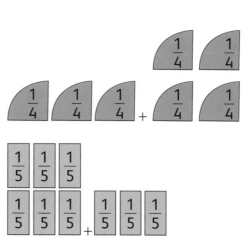

2 This diagram shows $2\frac{1}{3}$.

   a How many thirds is that in all?

   b What is $2\frac{1}{3} - 1\frac{2}{3}$?

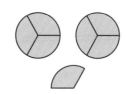

Unit 1 Add and Subtract Fractions and Mixed Numbers

Let's Think …

- How many quarters are there in $2\frac{1}{4}$?

- How many tenths are there in $1\frac{8}{10}$?

- How would you add $2\frac{1}{4}$ and $1\frac{8}{10}$?

To add or subtract *fractions*, the denominators need to be the same.

$$\frac{6}{8}+\frac{1}{8}=\frac{7}{8} \qquad \frac{7}{10}-\frac{3}{10}=\frac{4}{10}$$

When the denominators are different you can convert them to *equivalent fractions* that have the same denominator. You may need to *regroup* the answer.

$$\frac{3}{4}+\frac{1}{2}=\frac{3}{4}+\frac{2}{4}=\frac{5}{4}$$

$\frac{5}{4}$ means five quarters. This is equivalent to 1 whole $\left(\frac{4}{4}\right)$ and $\frac{1}{4}$.

You can regroup $\frac{5}{4}$ to make the *mixed number* $1\frac{1}{4}$.

When you work with mixed numbers you can regroup them to make your calculations easier.

Example 1

$4\frac{1}{4}-\frac{3}{4}$ Rename $4\frac{1}{4}$ to make more

quarters. Remember $1=\frac{4}{4}$.

$4\frac{1}{4}=3+\frac{4}{4}+\frac{1}{4}=3\frac{5}{4}$

$3\frac{5}{4}-\frac{3}{4}=3\frac{2}{4}$ Simplify the fraction: $\frac{2}{4}=\frac{1}{2}$

$\qquad\qquad =3\frac{1}{2}$

Example 2

$1\frac{7}{8}+\frac{3}{8}$ Add the fraction parts.

$=1\frac{10}{8}$ Regroup $\frac{10}{8}$ to get $\frac{8}{8}+\frac{2}{8}=1\frac{2}{8}$

$=2\frac{2}{8}$ Simplify $\frac{2}{8}$: $\frac{2}{8}\div\frac{2}{2}=\frac{1}{4}$

$=2\frac{1}{4}$

Example 3

$1\frac{3}{4}+4\frac{13}{20}$ Convert $\frac{3}{4}$ to an equivalent

fraction with a denominator of 20:

$$\frac{3}{4}\times\frac{5}{5}=\frac{15}{20}$$

$=1\frac{15}{20}+4\frac{13}{20}$

$=5+\frac{28}{20}$ Convert $\frac{28}{20}$ to an equivalent

mixed number: $\frac{20}{20}+\frac{8}{20}=1\frac{8}{20}$

$=5+1\frac{8}{20}$

$=6\frac{2}{5}$ Simplify $\frac{8}{20}$: $\frac{8}{20}\div\frac{4}{4}=\frac{2}{5}$

1 Do these calculations mentally. Write the answers only. Reduce to its lowest form.

a $\dfrac{7}{8} - \dfrac{4}{8}$

b $\dfrac{3}{4} - \dfrac{2}{4}$

c $\dfrac{5}{6} - \dfrac{4}{6}$

d $\dfrac{9}{7} - \dfrac{3}{7}$

e $1\dfrac{3}{5} - \dfrac{3}{5}$

f $2\dfrac{1}{3} + 3\dfrac{1}{3}$

g $4\dfrac{3}{10} + \dfrac{2}{10}$

h $5\dfrac{5}{9} + 3\dfrac{2}{9}$

i $1\dfrac{9}{10} + \dfrac{1}{10}$

2 Use equivalent fractions to find the answers to each calculation.

a $\dfrac{4}{5} + \dfrac{2}{3}$

b $\dfrac{7}{4} + \dfrac{4}{3}$

c $\dfrac{12}{7} - \dfrac{13}{14}$

d $\dfrac{1}{2} - \dfrac{2}{5}$

e $\dfrac{3}{10} + \dfrac{4}{5}$

f $\dfrac{7}{3} + \dfrac{3}{2}$

g $\dfrac{2}{6} + \dfrac{3}{9}$

h $\dfrac{8}{3} - \dfrac{5}{6}$

i $2\dfrac{2}{15} - \dfrac{4}{3}$

3 Calculate.

a $1\dfrac{1}{4} + 1\dfrac{1}{2}$

b $\dfrac{1}{5} + \dfrac{2}{3}$

c $\dfrac{4}{5} - \dfrac{1}{2}$

d $4\dfrac{7}{12} - 3\dfrac{3}{8}$

e $2\dfrac{1}{3} + 2\dfrac{4}{7}$

f $6\dfrac{3}{4} - 3\dfrac{5}{6}$

g $\dfrac{7}{5} + 2\dfrac{3}{5}$

h $\dfrac{5}{6} - \dfrac{1}{4}$

i $1\dfrac{5}{11} - \dfrac{10}{11}$

j $4\dfrac{1}{6} - 3\dfrac{4}{9}$

k $5\dfrac{3}{8} - 3\dfrac{1}{12}$

l $6\dfrac{7}{8} + 3\dfrac{3}{4}$

m $6\dfrac{4}{5} - 3\dfrac{2}{8}$

n $2\dfrac{1}{4} + 8\dfrac{2}{5}$

o $4\dfrac{1}{6} + 5\dfrac{3}{4}$

p $4\dfrac{7}{12} - \dfrac{12}{12}$

q $\dfrac{6}{13} + \dfrac{1}{2}$

r $8\dfrac{4}{5} - 6\dfrac{3}{10}$

4 Calculate. Give your answers as mixed numbers.

a $2\dfrac{1}{4} + 3\dfrac{1}{5} - 2\dfrac{1}{3}$

b $3\dfrac{1}{3} + 2\dfrac{3}{4} - 1\dfrac{2}{5}$

c $2\dfrac{1}{2} - \dfrac{2}{5} + 3\dfrac{3}{10}$

d $4\dfrac{2}{3} - 2\dfrac{1}{10} + \dfrac{3}{4}$

e $10\dfrac{2}{25} - 6 - 1\dfrac{1}{2}$

f $4\dfrac{13}{20} - 1 - 3\dfrac{49}{50}$

Looking Back

1 Write each of these fractions as mixed numbers.

a $\dfrac{8}{5}$

b $1\dfrac{1}{5}$

c $\dfrac{18}{5}$

2 Rewrite each mixed number as an equivalent number of fifths.

a $2\dfrac{3}{5}$

b $1\dfrac{1}{5}$

c $4\dfrac{3}{5}$

Unit 2 Add and Subtract Decimals

Let's Think …

Kim and Sam kept some tadpoles in a tank for their science project. They sketched the tadpoles and recorded the length of the body and tail (in cm) as they grew.

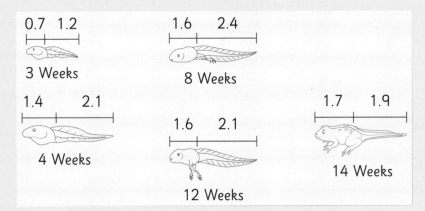

Use their notes to work out the total length of the tadpoles (in cm) at each stage of their growth. Tell your partner how you worked out the answers.

Decimals can be added and subtracted in columns using place value in the same way as whole numbers.

When you write decimals in columns to add or subtract them, you must line up the decimal point in the calculation and the answer.

You can fill empty places with 0 placeholders to make it easier to add or subtract in columns.

Example 1   Add $0.4 + 2.05 + 5.007$

0.4**00**	Write the decimals in columns. Line up the decimal points.
2.05**0**	Fill in 0 as a placeholder in empty places.
+ 5.007	Add in the same way as whole numbers.
7.457	Write the decimal point in the answer as well.

Example 2   Add $2.07 + 14.59$

₁	
2.07	Write the decimals in columns. Line up the decimal points.
+ 14.59	Regroup numbers as necessary.
16.66	Write the decimal point in the answer as well.

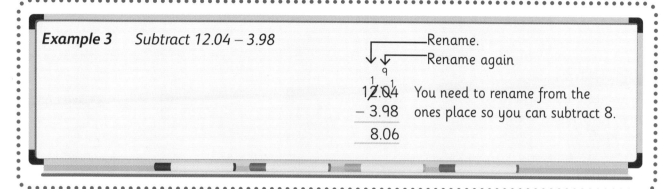

Example 3 Subtract 12.04 − 3.98

Rename.
Rename again

$$\begin{array}{r} 1 9 1 \\ 1\,2.\,0\,4 \\ -\ 3.\,9\,8 \\ \hline 8.\,0\,6 \end{array}$$

You need to rename from the ones place so you can subtract 8.

1 Add.

  a 0.4 + 0.37   b 1 + 0.897   c 0.1 + 2.39   d 4.601 + 0.22   e 0.3 + 13   f 21.3 + 34.24

2 Write in columns and add.

  a 18.22 + 71.666   b 0.47 + 13.3 + 8   c 17.54 + 11 + 0.7

  d 7 + 6.342 + 5.09   e 72.1 + 82.45 + 23.124   f 420.02 + 3.876 + 0.2

3 Subtract.

  a 65.47 − 13.25   b 14.26 − 8.01   c 59 − 36.05   d 1.75 − 0.6   e 1.76 − 0.998   f 1.75 − 0.356

> Estimating is important when you add or subtract decimals as it will give you some idea of the size of a reasonable answer.

4 Round each decimal to the nearest whole number and estimate the answer before doing each of these calculations.

  a 3.63 + 9.8 + 6.21   b 14.3 + 6.7 + 9.69   c 98.76 − 54.12

  d 18.23 − 10.15   e 0.57 + 0.66 − 1 + 0.92 − 0.03   f 64.37 − 24.39 + 38.5

5 Peter is 1.84 metres tall. His sister is 1.6 m tall. How much taller is Peter?

6 A taxi travels 23.47 km on Monday, 38.05 km on Tuesday and 29 km on Wednesday. What is the total distance it travelled?

Looking Back

Mrs Joyner asked a group of students to add 0.43, 12.084 and 3.8.
Which student has set out the work correctly?
What have the others done incorrectly?

Tamaya	Joshua		Kaylene	James	Linda
0.43	0.4300		043	0.43	0.430
12.084	12.084		12084	12.084	12.084
+ 3.8	2.7000		+ 38	3.8	3.800

Unit 3 Mixed Problems

Let's Think ...

Mr Bantam buys five items at the shops. The prices are given on the right.

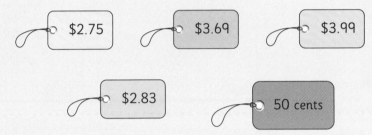

- How much change would he get from $20.00?

- How would you solve this problem?

- Try your method. Compare answers with a partner.

Problems involving fractions and decimals are very common in everyday life.

Just as with whole number problems, you have to read the problem carefully and look for clues that tell you whether to add, subtract or both.

Work with a partner. Read each problem through and talk about what you would need to do to solve it. Then, work on your own to work out the solutions.

1 To get to school, James walks $\frac{1}{2}$ km to the bus stop and then he takes a bus for 5.89 km to the school gates. How far does he travel in all?

2 Anna has a mass of 63 kg. Joshua has a mass of 57.32 kg and Shawn has a mass of $74\frac{1}{4}$ kg. What is their total mass?

3 What is the difference between 36.387 litres and 89.205 litres?

4 Nadia ate $\frac{1}{6}$ of a cake and Pete ate $\frac{1}{2}$ of the same cake. How much cake was left?

5 Shaundra went to the beach for $1\frac{3}{4}$ hours on Saturday and $2\frac{3}{5}$ hours on Sunday. How much time did she spend at the beach in total?

6 Jessica ran $2\frac{1}{2}$ km, then walked for $1\frac{3}{4}$ kilometres and then ran again for $\frac{3}{4}$ km. How far did she go altogether?

7 Ben watched TV for $3\frac{1}{3}$ hours and did homework for $1\frac{1}{4}$ hours.
   a How much time did he spend on both activities?
   b How much less time did he spend on his homework than on watching TV?

8 What is the perimeter of this shape?

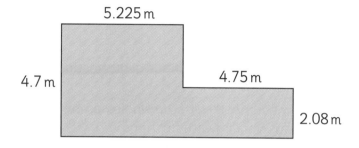

5.225 m

4.7 m

4.75 m

2.08 m

9 Juanita pours 1.045 litres of juice from a full 2 litre container. How much juice is left?

10 Nathan puts 450 g and 0.987 kg into a pan and then he measures the mass on a digital scale. Read the mass on the scale and work out the mass of the pan.

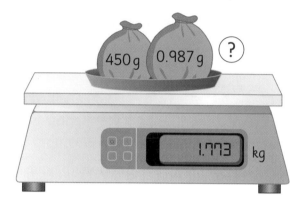

450 g   0.987 g   (?)

1.773   kg

11 The distance meter on Mr Jones' car showed the reading on Monday (on the left) and the reading on Friday (on the right). How many kilometres did he travel in this time?

MONDAY

| 2 | 3 | 4 | • | 8 |

FRIDAY

| 5 | 1 | 0 | • | 3 |

12 A large packet of flour contains 2.25 kg of flour when it is full. If there are 1 485 grams of flour left in the bag, how much has been used?

13 Mindy has a laptop and a printer next to each other on her desk. The desk is 1.2 m wide. The laptop is 32.5 cm wide and the printer is 0.418 m wide.

Does she have enough space to put a 50 cm wide box next to them?

Looking Back

Micah bought three items costing $2.62, $1.89 and $2.00. How much change would he get if he paid with a $10.00 bill?

Topic Review

What Did You Learn?

- If fractions have different denominators, use equivalent fractions to make the denominators the same so that you can work with them.
- To add or subtract mixed numbers, work with the whole numbers first then add or subtract the fractions using the rules for working with fractions.
- You can regroup mixed numbers to make equivalent fractions when you add or subtract.
- To add or subtract decimals, write them in columns with the same places above each other. The decimal points must line up in the calculation and the answer.
- You can fill in zeros as placeholders when you add or subtract decimals to make it easier to work with the columns.
- To solve problems involving fractions or decimals, work in the same way as with other problems.

Talking Mathematics

- What did you find easiest in this topic? Why?
- What did you find most challenging? Why?
- What three things are important when you solve problems involving fractions or decimals? Why?

Quick Check

1 What fraction is:
   a $\frac{1}{4}$ more than $\frac{1}{3}$?
   b $\frac{1}{2}$ less than $\frac{5}{8}$?
   c $1\frac{1}{2}$ greater than $2\frac{1}{6}$?
   d $2\frac{1}{4}$ smaller than $3\frac{1}{9}$?

2 Calculate.
   a $23.47 + 38.43 + 13$
   b $12.09 + 14.765$
   c $143.09 - 14.245$
   d $35 - 19.99$

3 An empty bucket weighs 1.025 kilograms. When it is filled with water it weighs 2.5 kilograms. What is the mass of the water.

4 1.025 litres of water is poured from a $2\frac{1}{2}$ litre container. How much water is left?

5 Make up three multi-step addition and subtraction problems involving money amounts. Swap with a partner and solve each other's problems.

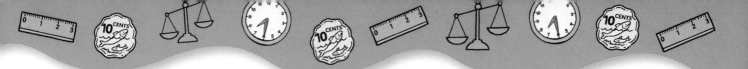

Topic 21 Capacity and Volume Workbook pages 65–67

Key Words

capacity

volume

cubic

container

millilitre

litre

pint

quart

gallon

▲ Which jars take up the most space? Which take up the least space?
Which hold the most?

Different containers can hold different amounts. When you measure a 3-D container, you can measure how much space it takes up. This is called the **volume**. You can also measure how much the **container** holds. This is its **capacity**. In this topic, you will estimate and measure volume using **cubic** units. You will measure capacity using **litres** and **millilitres**. You will also convert between units and solve problems involving volume and capacity.

Getting Started

1 Look at the containers in the picture above.

● Do the tallest containers always hold more? Why or why not?

● Which containers are easiest to pack? Why?

● Which containers take up the most space?

● Why do we need bottles and jars of different sizes?

2 Think about your own home.

a Which container in your home holds the most liquid?

b What containers hold more than 2 litres?

c What is the smallest container of liquid? What shape is it?

Unit 1 Measuring Volume

Let's Think …

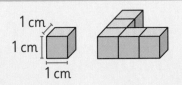

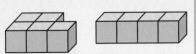

- Which of these objects shows 1 cm³?
- How much space does each other object take up? (Count the blocks.)
- Which is the longest shape?
- Which two objects take up the same amount of space?
- Does the longest shape take up the most space? Why or why not?

> *The amount of space that a solid object takes up is called its volume. You measure volume in cubic units. One cubic centimetre (1 cm³) has a height, width and depth of 1 cm. You measured volume in Grade 4, using cubic centimetres.*

1 Count the blocks of each shape to work out its volume.

a 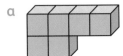   b   c 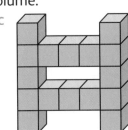   d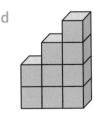

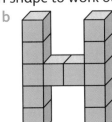

> *Look again at the shape **d** above. The block has the following dimensions:*
>
> length = 4 blocks   width = 3 blocks   height = 2 blocks
>
> *Instead of counting up the total blocks, you can use this formula to calculate the volume:*
>
> volume = length × width × height
>
>   = 4 × 3 × 2 = 24
>
> *The volume of the shape is 24 cm³.*

2 Use the formula to calculate the volume of
   a a box with height of 8 cm, width of 5 cm and height of 1 cm
   b a box with height of 15 cm, width of 10 cm and height of 3 cm
   c a cube with all dimensions equal to 12 cm.

Looking Back

Collect some boxes from products such as cookies, cereals or any other dry groceries. Work in pairs or groups.

a Estimate the dimensions of each box.
c Measure the box and calculate the actual volume.

b Use your estimated dimensions to estimate the volume.
d Record your estimates and measurements in a table.

Unit 2 Measuring Capacity

1 litre   500 mL

Capacity is a measure of how much a container can hold. You measure capacity in litres and millilitres.

1 litre (L) = 1 000 millilitres (mL)

To convert litres to millilitres, multiply by 1 000.

$$4\,L = \boxed{}\,mL$$

$$4 \times 1\,000 = 4\,000\,mL$$

To convert millilitres to litres, divide by 1 000.

$$175\,mL = \boxed{}\,L$$

$$175 \div 1\,000 = 0.175\,L$$

1 a Collect a range of bottles, cups and jars.

   b Arrange them in order, from those that look like they hold the least to those that hold the most. Give each container a letter: A, B, C, D and so on.

   c Write the capacity (in mL or L) of each container. Measure it using water if it does not have a label.

   d Was your order in part **b** correct?

   e Convert the capacities so you have recorded them in both millilitres and litres.

   f Copy and complete a table like this:

Container	Estimated Capacity	Actual Capacity (mL)	Actual Capacity (L)
A			
B			
C			
D			
E			
F			

You use many other units of capacity in the kitchen.

Cups and Spoons

1 teaspoon (tsp) = 5 mL

1 tablespoon (tbsp) = 15 mL

1 metric cup = 250 mL

$\frac{1}{2}$ cup = 125 mL

$\frac{1}{4}$ cup = 62.5 mL (usually rounded to 60 mL)

Customary Measures

Sometimes a recipe may use the older imperial system of measures, or customary measures, which uses cups, pints, fluid ounces, quarts and gallons.

1 cup = 8 fluid ounces (fl oz)

1 pint (pt) = 2 cups

1 quart (qt) = 2 pints

1 gallon (gal) = 4 quarts

2 Read the recipe for this sauce. Write the measurement for each ingredient in millilitres.

3 Work these out.

a $\frac{1}{2}$ cup = ☐ fl oz

b $2\frac{1}{2}$ quarts = ☐ pints

c $5\frac{1}{2}$ gallons = ☐ pints

d 1 quart = ☐ gallons

4 Which customary measure is roughly equal to one litre?

5 At home or at a store, find bottles with capacity given in customary measures.
Make a list of five products and the capacities of their containers.

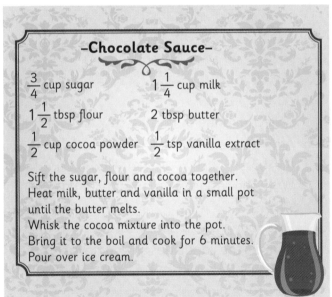

—Chocolate Sauce—

$\frac{3}{4}$ cup sugar $1\frac{1}{4}$ cup milk

$1\frac{1}{2}$ tbsp flour   2 tbsp butter

$\frac{1}{2}$ cup cocoa powder $\frac{1}{2}$ tsp vanilla extract

Sift the sugar, flour and cocoa together.
Heat milk, butter and vanilla in a small pot until the butter melts.
Whisk the cocoa mixture into the pot.
Bring it to the boil and cook for 6 minutes.
Pour over ice cream.

6 a Mila needs a gallon of milk to make pancakes for a school fair. The store only has milk in containers of one pint. How many must she buy?

b Noah buys 8 litres of juice for a party. There will be 20 guests, and he wants to make sure there is enough for each guest to have at least two cups of juice. How much more juice must he buy?

c A chef makes 1.5 litres of apple sauce to use in apple pies. Each pie needs $\frac{1}{2}$ cup of apple sauce. How many pies can she make altogether with the apple sauce?

7 Look at this recipe.
   a Express all the quantities in mL.
   b Calculate how much you would need of
      each ingredient in order to make double
      the quantity.

8 Write three questions of your own involving
   measuring capacity. In groups, put together
   your questions to make a booklet of problems
   to solve. Exchange booklets with another
   group and solve each other's questions.

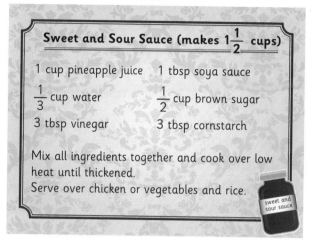

Sweet and Sour Sauce (makes 1$\frac{1}{2}$ cups)

1 cup pineapple juice   1 tbsp soya sauce

$\frac{1}{3}$ cup water $\frac{1}{2}$ cup brown sugar

3 tbsp vinegar   3 tbsp cornstarch

Mix all ingredients together and cook over low
heat until thickened.
Serve over chicken or vegetables and rice.

sweet and
sour sauce

Looking Back

1 Write the customary unit that is equal to:
   a 8 fl oz   **b** 2 cups
   c 2 pints   **d** 8 pints.

2 How many litres is:
   a 24 cups   **b** 4 cups
   c 1 cup   **d** 3 cups.

3 Write the following in litres.
   a 5 mL   **b** 159 mL
   c 4579 mL   **d** 13 244 mL

4 Which of the three jars would best fit this
   curry mixture?
   5 tbsp ground coriander,
   2 tbsp ground cumin,
   1 tbsp ground turmeric,
   2 tsp ground ginger,
   2 tsp dry mustard,

1$\frac{1}{2}$ tsp black pepper,

1 tsp cinnamon,

$\frac{1}{2}$ tsp cloves,

$\frac{1}{2}$ tsp ground cardamom,

$\frac{1}{2}$ tsp ground chilli peppers.

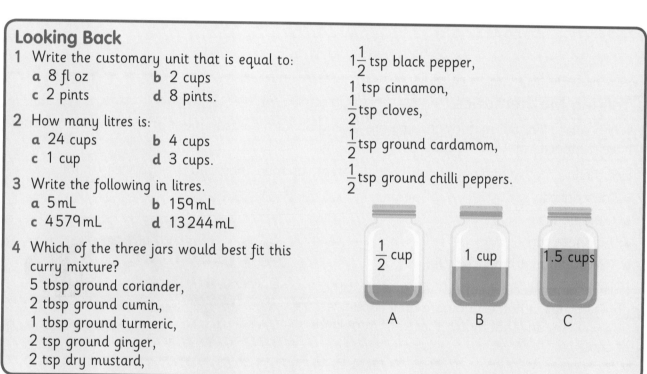

$\frac{1}{2}$ cup   1 cup   1.5 cups

A   B   C

Topic Review

What Did You Learn?

- The measure of space that a solid object takes up is called volume. You measure volume in cubic units such as cubic centimetres (cm^3).
- You can calculate volume by adding up cubes.
- For cuboids, you can multiply the length × width × height to find the volume.
- Capacity is the amount of liquid a container can hold.
- You measure capacity in litres and millilitres.
- A cup is 250 mL. 4 cups = 1 litre.
- In the old imperial system of measures, you use cups, fluid ounces, pints, quarts and gallons.

Talking Mathematics

When you talk about metric cups, you are talking about a measurement of 250 mL. Not all cups have a capacity of 1 metric cup!

Use the internet to find out about these different cup measures, and their capacity in mL.

- Imperial cup
- Canadian cup
- Japanese cup
- Gō (traditional Japanese cup)

Collect some tea cups, coffee cups and any other drinking cups at home. Compare their capacities. Can you find any that match these different kinds of cups?

Quick Check

1 Write the correct term for each definition.
   a The amount of space an object takes up, measured in cubic units.
   b The amount of liquid a container can hold, measured in mL and L.

2 How many…
   a millilitres are in half a litre?
   b cups are in 2 litres?
   c pints are in a quart?

Topic 22 Transformations Workbook pages 68–70

☐ These three logos were made using different types of transformation.
Can you identify translation, reflection and rotation?

To **transform** something means to change or alter it. In geometry, the transformation of figures also involves change. When you **translate**, **reflect** or **rotate** a figure, you move it in some way. In this topic, you will revise and develop your understanding of these three transformations of plane figures. You will also learn what it means for two figures to be **congruent**.

Getting Started

1 Find pictures of logos of well-known companies (such as car manufacturers) in magazines, newspapers or on the internet. See if you can find at least three examples of a logo that has translation, reflection or rotation in its design.

2 **A B C D E F**

U V W X Y Z

a Which of these can be examples of reflection if you draw in a mirror line?

b How many times will each letter fit onto itself if you rotate it through 360 degrees?

3 If you look at letters in a mirror, they appear backwards. Try to write your name as it would appear in a mirror. Use a mirror to look at what you have written. If you did it correctly, you should see your name clearly in the mirror.

Unit 1 Translation, Reflection and Rotation

When you move a figure in a straight line in any direction, you translate it to a new position.

The figure stays the same size and does not turn in any way, but slides to a new position.

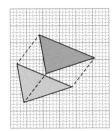

When you flip a figure over a line you reflect it. The line of reflection is also called a mirror line and can be horizontal, vertical or diagonal.

The figure stays the same size. You end up with a mirror image of the original figure.

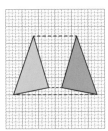

When you turn a figure around a fixed centre point, you rotate it.

The figure remains the same shape and size, but faces a different direction as it turns. When a figure has made a full rotation, it faces the same way as the original.

The rotation can be clockwise or anti-clockwise.

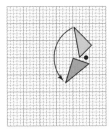

Two figures are congruent when they have exactly the same shape and size. If you put the two figures on top of each other, they will match exactly. When a figure is translated, reflected or rotated, it still has the same shape and size as the original, so it will be congruent with the original figure. Two figures that are not congruent or the same, are incongruent.

These figures are all congruent, as they all have the same shape and size even though the shapes have moved in some way.

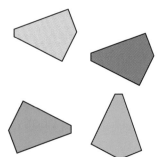

1 Name the following transformations.

a   b   c   d   e

Musical notes   Footprints   Propeller   Basketball   Dartboard

2 Pick the correct answer.

                                        A   B   C

a   Which diagram shows a translation of the rectangle?

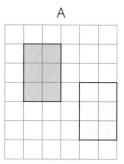   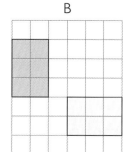 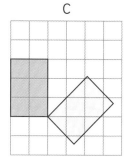

b   Which diagram shows a reflection of the triangle?

                                        A   B   C

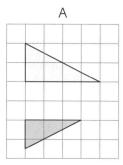   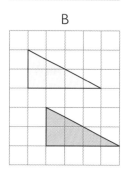 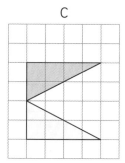

c   Which diagram shows the square being rotated 180° clockwise around the marked point?

                                        A   B   C

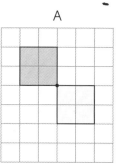   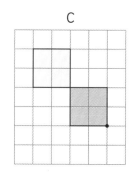

3 State whether the following figures are congruent or not.

    A   B   C   D   E

> **Looking Back**
>
> Write a letter to a friend in which you explain to him or her in your own words what translation, reflection and rotation are. Also explain what it means when you say that two figures are congruent. Remember to include examples in your letter!

Topic Review

What Did You Learn?

- A translation slides a figure to a new position in a straight line in any direction. The figure remains the same, it is just in a different position.
- A reflection flips a figure across a mirror line to a new position. The figure is now a mirror image of the original.
- A rotation is a turn around a fixed point. The figure faces in a different direction, depending on the amount of rotation.
- Congruent figures are identical but may be moved or turned, for example through translation, reflection or rotation.

Talking Mathematics

Give the mathematical word for each of the following.

a Two figures that are exactly the same but not necessarily facing the same direction.

b A turn that moves a figure around a fixed centre point.

c The transformation that flips a figure over a line of reflection.

d A move that slides a figure in a straight line to a new position.

Quick Check

Can you identify any transformations that were applied to create this design?

Topic 23 Perimeter and Area Workbook pages 71–72

Key Words

dimensions
area
perimeter
rectangles
squares
square units
square metre (m²)
square centimetre (cm²)
square kilometre (km²)
circumference
diameter
radius
compass
pair of compasses

▲ This is a bird's-eye-view of a soccer field. What shapes can you see on the field? The whole area is covered with grass. The white lines are painted. How do you know how much grass to plant? How do you know how much paint to use for the lines?

In real life, we often need to work out the **dimensions** of flat shapes, such as a soccer field, a tennis court or the floor of a room. In order to work out these dimensions, we may need to calculate **area**: how much flat space the figure takes up, or **perimeter**: the distance around the edge of the **figure**. In this topic, you will estimate and measure perimeters of objects, identify and describe the circumference of circles, calculate areas of **rectangles** and **squares** using **square units**, and learn when to use area calculations and when to use perimeter.

Getting Started

1 Use a piece of paper. Cut out a square that is 10 cm long and 10 cm wide. This is one square. How many of these squares would you need to cover the area of:

- your desk
- the whiteboard or blackboard
- the window in your classroom?

2 For each item above (desk, board, window), guess the distance around the edge of it. Then use a ruler or tape measure to measure it accurately. How accurate was your guess?

Unit 1 Measuring Perimeter

Let's Think …

- How would you measure the distance around the square and the rectangle?
- How would you measure the distance around the circle and the oval?

The perimeter of an object is the distance around it. You can guess the approximate measurement, without using any measuring instruments. This is called estimating.

To measure the actual perimeter, you need to measure the length of each side with a ruler or tape measure. Then add up the sides.

If the lengths of the sides (or dimensions) are given, just add them up. For curved sides, you can lay a piece of string along the line and then measure the string.

1 Estimate, then measure, the perimeters of the following shapes.

  a The light switch in your classroom.

  b Your mathematics textbook.

  c Your homework diary.

  d The door of your classroom.

2 Work in pairs or groups. Your teacher will assign each group a different place in your school, such as a classroom, a field, a quadrangle or a yard. Estimate and then measure the perimeter of each place.

3 Find the perimeter of each shape without measuring.

a

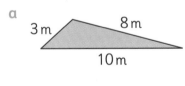

3 m 8 m 10 m

b

15 m 15 m 11 m 14 m 12 m

c

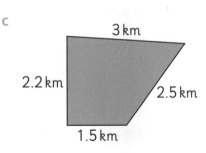

3 km 2.2 km 2.5 km 1.5 km

4 Use the measurements given on these sketches. Calculate the perimeter of each rectangle.

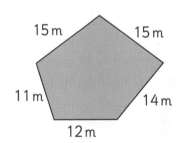

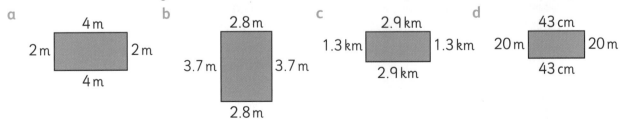

a 4 m 2 m 2 m 4 m

b 2.8 m 3.7 m 3.7 m 2.8 m

c 2.9 km 1.3 km 1.3 km 2.9 km

d 43 cm 20 m 20 m 43 cm

5 Look at your work from question 4.

    a Can you suggest a shortcut for calculating the perimeter of a rectangle?

    b A square has four equal sides. Can you work out a rule for calculating the perimeter of a square?

In a rectangle, each side is equal in length to its opposite side.

To calculate the perimeter of a rectangle, you can add up the four sides.

Or you can use this formula:

(2 × length) + (2 × width)

A square has four equal sides.

To calculate the perimeter of a square, you can add up the four sides.

Or you can use this formula:

4 × length of one side

6 Work out these perimeters.

    a A square cake with each side 20 cm in length.

    b A pillow with a length of 60 cm and width of 45 cm.

    c A rectangular window with length 2.4 m and height 0.5 m.

7 Solve these problems.

    a A poster board has a length of 90 cm and width of 65 cm. Julia needs enough tape to make a border all the way along the edge of the poster. Calculate the perimeter of the board in order to work out how much tape she needs.

    b Alicia makes a square cake with each side 24 cm in length. What length of ribbon would she need to go around the cake?

    c A photograph is 6 inches wide by 8 inches long. James works out the perimeter plus an extra half-inch on each end of each side, in order to work out how much tape he needs to stick the photograph in his album. What is the total length of the tape?

    d The perimeter of a square cushion is 160 cm. What is the length of each side?

Looking Back

1 Write your own definition of perimeter.

2 Write the formula for calculating:

    a the perimeter of a rectangle

    b the perimeter of a square.

3 Calculate the perimeter of each item.

    a a photo album with height of 21.5 cm and width of 18.5 cm

    b a square sticker with all sides equal to 24 mm

    c a birthday card with height of 155 mm and width of 87 mm.

Unit 2 Measuring Area

Let's Think …

The block in the picture shows one square centimetre (1 cm²).

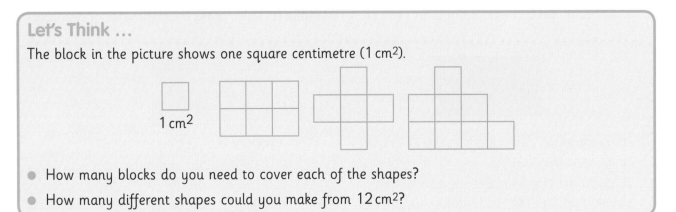

1 cm²

- How many blocks do you need to cover each of the shapes?
- How many different shapes could you make from 12 cm²?

> *Area is a measure of how much space a flat shape takes up. You measure area in square units such as cm². One square centimetre (1 cm²) is a square with a length and width of 1 cm. One square metre is a square with a length and width of 1 m.*

1 Use centimetre squared paper. Draw the following shapes.

  a A square with an area of 9 cm².

  b A square with an area of 16 cm².

  c A rectangle with an area of 8 cm².

  d A rectangle with an area of 15 cm².

2 a Work out the length of each side of the centimetres and rectangles that you drew in question 1.

  b Can you work out the relationship between the lengths of the sides of a rectangle and its area?

You can calculate the area of a rectangle using this formula:

Area = length × width

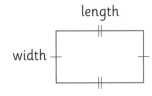

Remember, for a square, the length is equal to the width, so the formula is:

Area = length × length

Or Area = (length)²

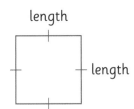

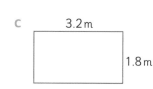
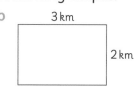
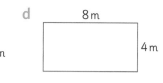

3 Calculate the area of the following shapes.

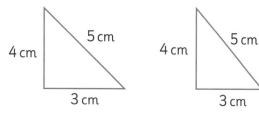

a
9 m

b
3 km
2 km

c
3.2 m
1.8 m

d
8 m
4 m

  a a square with each side equal to 9 m

  b a piece of land with length equal to 3 km and width equal to 2 km

  c a wall with height of 1.8 m and width of 3.2 m

  d a billboard with width of 8 m and height of 4 m

4 Work with a partner. Draw accurate versions of these two triangles. Tape your triangles together to create a rectangle.

5 cm
4 cm
3 cm

5 cm
4 cm
3 cm

  a Work out the area of the rectangle you made.

  b What is the area of each triangle? (Hint: You doubled the triangles to make the rectangles).

5 Do these calculations involve perimeter or area?

  a Working out the amount of wallpaper you need to cover a wall.

  b Finding how much fencing is needed to go around a field.

  c Calculating how much wood to use for a picture frame.

  d Working out how many patches of roll-on grass are needed to fill a garden bed.

6 Use centimetre squared paper. Make different shapes that have:

  a an area of 24 cm²   b a perimeter of 8 cm.

7 TRUE or FALSE? Use what you did in question 6 to help you explain why.

  a Two shapes can have the same area but different perimeter.

  b Two shapes can have the same perimeter but different area.

  c A shape can have the same perimeter as its area.

Looking Back

1 Which of the following units is NOT used for calculating area:

         cm²   km²   cm   mm²

2 What is the difference between perimeter and area?

3 A garden has a length of 3.5 m and a width of 2.2 m.

  a What is its perimeter?   b What is its area?

Unit 3 Working with Circles

Let's Think …

How can you measure the perimeter of a circle?
Can you use square centimetres to measure the
area it takes up?
Why is measuring the perimeter and area of circles
more difficult than measuring rectangles and
squares?

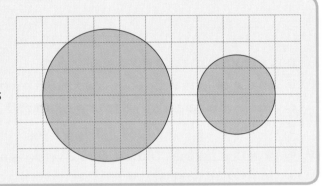

The perimeter of a circle is called the circumference.
The distance from the centre point of the circle to any
point on the circumference is called the radius.
A diameter is any straight line that passes from one
point on the circumference, through the centre point,
to the circumference on the other side.

The length of the circumference is always twice the
length of the radius. d = 2r.

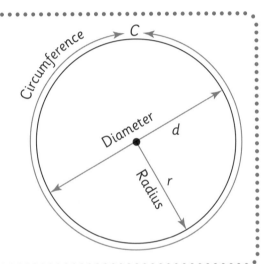

1 You will need:

- some round objects to trace around
- paper
- string
- tape measure
- scissors
- calculator

a Trace around three different round objects, and cut out the circles.

b Number them 1 to 3.

c Fold each circle in half to find the diameter. Measure the length of the diameter.

d Use string to measure the circumference of each circle.

e Copy and complete the table below. Use a calculator to help you work out what you get if you
divide the circumference by the diameter for each circle.

	Diameter	Circumference	Circumference ÷ Diameter
1			
2			
3			

f Compare your findings with the rest of the class. What do you notice?

Pi is a special number that you write like this: π

Mathematicians discovered pi by studying the ratio or relationship between the circumference of a circle and its diameter.

$$\pi = 3.14 \quad or \quad \frac{22}{7}$$

When you work with circles, you can use the following formula for calculating the circumference of a circle:

circumference = π × diameter

or $C = \pi \times d$

Remember that the diameter is always twice the length of the radius. So you can also write the formula like this:

circumference = 2 × π × radius

or $C = 2 \times \pi \times r$

A *compass* (also called a *pair of compasses*) is a mathematical instrument for drawing circles. It has two arms: one with a pointed end, and one with a holder for a pencil. When you use a pair of compasses, the pointed end points to the centre of the circle. The pencil traces the circumference.

The distance between the pointed end and the pencil point form the radius of the circle. You can adjust the arms in order to determine the size of the circle you want.

2 Use a pair of compasses to draw a circle with a radius of:

 a 3 cm   b 5 cm   c 7.5 cm

3 Use a ruler to measure the diameter of each circle you drew in question 2. (Remember, the diameter must pass through the centre point.) Does it follow the rule that $d = 2 \times r$?

4 Use your compass to draw a pattern made up of repeated circles. Show your pattern to the rest of the class and explain how you made the pattern.

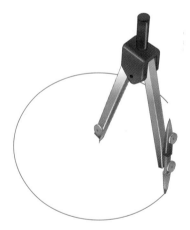

Looking Back

1 In your own words, describe the relationship between the radius and diameter of a circle.

2 What is another name for the perimeter of a circle?

3 Write pi as a decimal and as a fraction.

4 What is the formula for the circumference of a circle?

Topic Review

What Did You Learn?

- The perimeter is the distance around an object.
- You can calculate the perimeter by adding up the lengths of each side of a figure.
- The formula for perimeter of a rectangle is twice length plus twice width.
- The formula for perimeter of a square is four times the length of one side.
- The circumference of a circle is the distance around the circle or its perimeter.
  Pi $(\pi) = 3.14$ or $\dfrac{22}{7}$.
- The circumference C is equal to π times the diameter d or $C = \pi \times d$ or $C = 2 \times \pi \times r$
- Area is the amount of space a flat figure takes up, measured in square units.
- To calculate the area of a rectangle, $A = l \times w$

Talking Mathematics

1 Look up the meanings of these words in a dictionary: circumstance, circumnavigate, circus, circumvent. What do all these words have in common?

2 Look up the meanings of these words: radius, radiate, radio. In which ways does the meaning of 'radius' link to these other everyday words?

3 Explain in your own words the main difference between calculating the perimeter of a square and the perimeter of a rectangle?

Quick Check

1 Margaret has a piece of cardboard that is 85 cm long by 60 cm wide. She cuts out two squares of 20 cm × 20 cm. What area does she have left?

2 I draw a circle with a radius of 25 cm.

   a What is the diameter of the circle?

   b Would it be a fair guess that the circumference must be greater than 100 cm? Why?

Topic 24 Probability Workbook pages 73–76

▲ Imagine you roll a green die and a red die. In how many ways can you get a total of 2? In how many ways can you get a total of 7?

Most events cannot be predicted with absolute certainty. All you can do is to say how likely they are to happen, using the idea of **probability**. Probability gives you an indication of what is most or least expected to happen. In this topic, you will build on your previous knowledge of probability. You will learn how probability is expressed mathematically, and how you can use experiments to test probabilities. You will revisit ways of writing probability, and find out what **chance** is.

Getting Started
1 Work with a partner.
   a Take a coin and check which side is heads and which side is tails.
   b Now flip the coin 30 times and for each flip make a tally mark for either heads or tails.
   c Compare your results with those of the rest of the class. Use a table like this to tally your results.

Result Tally	
Heads	
Tails	

Unit 1 Chance, Outcomes and Probability

- What result did you expect when you flipped the coins?
- What do you think are the chances of the coin landing on heads?
- What about landing on tails?
- How does your result compare with your classmates' results?

We use the word 'chance' in everyday life in expressions like 'chances are', 'I met her by chance' or 'they don't stand a chance'.

In mathematics, chance refers to the probability that a random event will happen; for example, you might ask 'What is the chance of rolling a 5 with a six-sided die?'.

The outcome of rolling a die is random. You cannot predict what you will roll because any outcome from 1 to 6 is possible. However, you can work out the probability of each outcome.

The probability of an event is the likelihood that the event happens. You can describe this likelihood using words such as impossible, unlikely, likely and certain. You can also use numbers to express the probability that something happens.

You can write it like this, where P stands for the probability:

$$P = \frac{Number\ of\ favourable\ outcomes}{Number\ of\ possible\ outcomes}$$

Example 1

You have a bag with ten marbles in it. Three marbles are blue, five are red and two are green. You want a red marble, but you can only pull out one marble without looking into the bag.

What are your chances of pulling a red marble from the bag?

The favourable outcome is any red marble. There are five red marbles.

The number of possible outcomes is ten, since there are ten marbles in the bag.

Use the formula:

$$P(red\ marble) = \frac{Number\ of\ favourable\ outcomes}{Number\ of\ possible\ outcomes} = \frac{5}{10} = \frac{1}{2}$$

When you take a marble out of the bag, the chances that it will be red are 5 out of 10. You can simplify this to 1 out of 2. So, the probability of randomly pulling out a red marble can be expressed as the *fraction* $\frac{1}{2}$.

Example 2

Use the same bag of marbles as in Example 1. What is the probability of pulling out a purple marble from the bag?

$$P(\text{purple marble}) = \frac{\text{Number of favourable outcomes}}{\text{Number of possible outcomes}} = \frac{0}{10} = 0$$

There are no purple marbles in the bag so it is impossible to pull out a purple marble.

Example 3

What are the chances of pulling out an orange marble from a bag containing 5 orange marbles?

$$P(\text{orange marble}) = \frac{\text{Number of favourable outcomes}}{\text{Number of possible outcomes}} = \frac{5}{5} = 1$$

As all the marbles are orange, you are certain to pull out an orange marble.

The probability of an event can be expressed as a number from 0 to 1.

If an event is impossible, the probability of it happening is 0 and if an event is certain, the probability of it happening is 1.

All other probabilities will be a fraction somewhere between 0 and 1, depending on the particular circumstances.

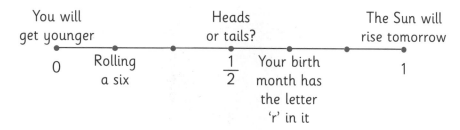

The probability line shows an impossible event at 0 and a certain event at 1. Can you determine the probability of the other two events as fractions and as ratios?

1 State whether the following are mostly games of chance or games of skill.

a A tennis match.

b A game of snakes and ladders.

c Spinning a spinner to determine a winning number.

d Matching the card memory games.

e Drawing a name out of a hat.

2 Name three situations that involve chance.

3 You are given a container holding seven pink balls, three yellow balls and two black balls. Write down each probability as a fraction when randomly taking one ball out of the container.

a P (pink)

b P (yellow)

c P (black)

d P (pink or yellow)

e P (not black)

f P (not purple)

4 Each of the letters of the word BAHAMAS is written on a separate piece of paper, folded and put in a hat. One piece of paper is randomly pulled out. What is the probability that the letter on the piece of paper is an A? Write your answer as a decimal.

5 This spinner is shaped like a regular pentagon with a toothpick through the centre. It is divided into five equal triangles, each with a number from 1 to 5 on it.

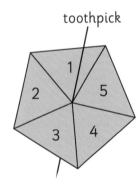

If you spin the spinner, what is the probability that it lands on an odd number?

Looking Back

Yohan flips a coin 100 times. What result would you expect?

The coin lands on heads 58 times and on tails 42 times.
What can you say about this result?

Topic Review

What Did You Learn?

- Chance is the possibility of a random event happening. Any one of a set of possible outcomes can take place.

- The probability of an event is the chance that it will happen. Probability can be expressed as a decimal.

- The probability of an event can be calculated using the following formula, where P is probability:

$$P = \frac{\text{Number of favourable outcomes}}{\text{Number of possible outcomes}}$$

- The probability of an impossible event is 0 and the probability of a certain event is 1. All other probabilities are fractions between 0 and 1.

Talking Mathematics

Pair the phrases on the left with the best match on the right.

When the outcome of an event is random   Experimental probability

The result of doing an experiment to test a probability   Fraction

A possible result of a probability experiment   Outcome

Probability of $\frac{7}{11}$ Chance

Quick Check

1 The table below shows data about randomly picking beads from a bag. What is the missing number? How many beads do you think there are in the bag?

Bead	Probability
Yellow	$\frac{2}{7}$
Black	?
Blue	$\frac{4}{7}$

2 If you flip a coin 4 times, what is the best prediction possible for the number of times it will land on heads?

3 Which of these results can occur when a coin is flipped 5 times?

a H, H, H, H, T

b H, T, H, T, H

c T, T, H, T, T

Topic 25 Looking Back

Revision A

1 Your teacher writes the following numbers on the board:

32 567 000   950 342

8 020 505   172 900 000

a Write each number in words.

b What is the value of the digit 2 in each number?

c Put the numbers in ascending order.

d Write the largest number in expanded notation.

2 Look at this pattern.

a Look only at the green triangles. Describe in words how the pattern works.

b Write a number sequence to match the pattern of green triangles.

c Work out how many green triangles there will be in the 10th shape without drawing it.

d Looking at all the triangles (green and white), write a number sequence to match the pattern.

e Describe the pattern of all the triangles and calculate how many triangles will be needed for the 8th shape.

3 Arrange each set of fractions in descending order.

a $\frac{1}{2}, \frac{7}{8}, \frac{3}{4}, \frac{3}{12}$ b $1\frac{1}{6}, 2\frac{1}{6}, 1\frac{3}{12}, 1\frac{8}{24}$

c $\frac{15}{25}, \frac{1}{5}, \frac{4}{10}, \frac{40}{50}$

4 Work out the following metric conversions.

a $3\,kg = \boxed{}\,g$

b $8\,550\,mg = \boxed{}\,g$

c $1.2\,t = \boxed{}\,kg$

5 The temperatures in different cities are shown in this table

City	Temperature
Moscow	−2 °C
Nassau	75.2 °F
London	7 °C
Miami	21 °C
Sydney	32 °C

a Describe in words how to convert temperatures from Fahrenheit to Celsius.

b Convert the temperature in Nassau from Fahrenheit to Celsius.

c Which is the coldest city and which is the hottest city?

d The temperature in Moscow rises by 5 °C. What is the temperature now?

e The temperature drops by 8 °C in London and Sydney. What are the temperatures in London and Sydney now?

6 Use $<$, $>$, or $=$ to compare the following numbers:

a 0.678 and 0.786

b $\frac{3}{4}$ and 0.8

c 0.875 and $\frac{7}{8}$

d 0.5 and 0.500

7 a Name the figure.

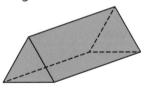

b How many faces, vertices and edges does it have?

c Draw a cylinder and label all faces, edges and vertices, if it has any, on the diagram.

Revision B

1 Sarah, Lee and Claudia decide to combine their pocket money and buy themselves a bike to share.

Sarah has $53.65, Lee has $48.10 and Claudia has $65.40.

a Round each value to the closest dollar.

b Estimate how much money they have in total.

c How much more money does Claudia have than Sarah?

d Round $53.65 to the nearest 10th of a dollar.

2 Look at the diagram.

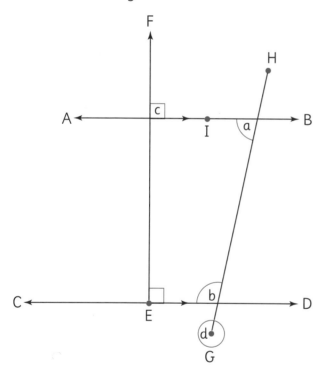

a Classify angles a, b, c and d.

b From the diagram, give examples of

i Parallel lines

ii Perpendicular lines

iii A point

iv A line segment

v A line

c Using a protractor, measure angle a and angle b.

3 You have these two sets of data:
A = {1, 4, 5, 7, 9}
B = {2, 3, 5, 6, 8}

What is the mean, median and range for each set?

4 Nick wants to make a small ladder. He needs two longer pieces of wood for the two side rails measuring 1.5 m each. He will have 7 steps in the ladder measuring 40 cm each. Nick has one long piece of wood measuring 65 dm.

a Does Nick have enough wood to make his ladder?

b Does he have any wood left over? If so, how much is left over?

5 Write the first 8 multiples of 3 and 5. What is the least common multiple?

6 Write down the factors of 8 and 20. Circle the common factors and indicate the greatest common factor.

7 Write 8.723 m in expanded form using different units.

8 Name the following polygons.

a

b
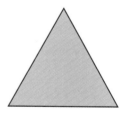

c
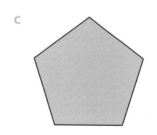

d
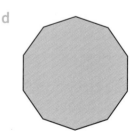

Revision C

1 Adam gets pocket money every week based on the chores he has completed at home. In July, Adam got $12.00, $8.50, $14.25, $5.25.

   a How much pocket money did Adam get in total in July?

   b Write $8.50 and $5.25 as mixed numbers with fractions.

   c Adam spends $\frac{1}{4}$ of his money on sweets and $\frac{3}{8}$ of his money on a game. How much money has he spent?

   d How much money does he have left?

   e Which is more: the amount Adam has left or $\frac{7}{16}$ of the total amount he started with?

2 John has built a tree house and wants to take his toys and games up into the tree house.

   He has a pulley system that can carry 3.5 kg at a time. If he has 15 kg of toys and games, how many times must he use the pulley system?

3 Calculate the following.

   a 4$\overline{)4321}$ b 6$\overline{)1962}$

   c 25$\overline{)725}$ d 12$\overline{)630}$

4 Jack buys 865 apples and would like to put them into bags of 15 apples.

   a How many full bags can he make?

   b How many apples are left over?

5 You do a survey with your friends looking at their favourite desserts. You use your data to draw the following graph.

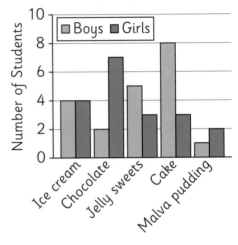

Favourite Desserts

   a Which dessert is the favourite for the boys? And the girls?

   b Which dessert do boys and girls like equally?

   c Is it TRUE or FALSE to say that Malva pudding is the least favourite for both boys and girls?

   d Is it TRUE or FALSE to say that the number of girls that like malva pudding best is the same as the number of boys that like chocolate best?

6 Write the following times in digital notation.

   a Twenty-five minutes to three in the afternoon.

   b Quarter past eleven in the morning.

   c Half past eight in the morning.

   d Midnight.

   e Now write your answers for parts **a** to **d** in 24-hour format.

7 Sarah decided to go for a run. She left her house at quarter past 10 in the morning and returned home at noon.

   a What was her starting time in digital notation?

   b Write her ending time in 24-hour format.

   c What was the duration of her run?

8 What is the volume of the following?

   a A box with a length of 12 cm, width of 10 cm and depth of 2 cm.

   b A box with a length of 0.1 m, width of 5 cm and depth of 3 cm.

9 Are the following figures congruent or incongruent?

   a

   b

   c

10 Sam is hosting a party and needs to buy cool drinks for his guests. He buys 20 L of cola and 12 L of sparkling water. He would like to make sure his guests have 2 cups of cola each and 1 cup of sparkling water. Does he have enough of each if he has 45 guests?

11 In a party bag, you get 5 chocolates, 3 lollipops, 8 jelly sweets and 2 bubble gums. You pick one sweet from the bag at random.

   a What is the probability that you will pick a jelly sweet?

   b Are you more likely to pick a lollipop or a bubble gum?

   c What is the probability that you will pick a fizzer?

12 Nicole buys this photo frame.

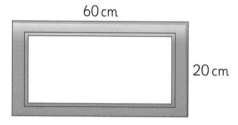

   a What is the perimeter of the frame?

   b What area will the frame take up on her wall?

13 In this drawing of a circle, the radius is 3 cm.

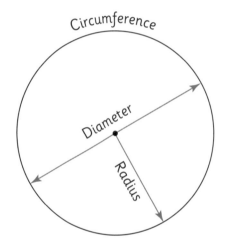

   a What is the formula for calculating the diameter? What is the diameter with radius 3 cm?

   b Give two formulae for calculating the circumference of a circle.

   c Using one formula, calculate the circumference of this circle.

14 Chantel has $22.25 and Laura has $41.75.

   a How much more money does Laura have than Chantel?

   b If Chantel gets another $7.75 and Laura gets another $8.25, how much do they each have?

Key Word Reference List

The key words that you learned this year are listed here in alphabetical order. If you cannot remember the meaning of a word, turn to the page number that is given next to the word. Read the definition and look at the pictures or examples to help you remember what the word means.

accuracy (page 75)

acute angle (page 68)

add (pages 79, 112)

approximate (page 54)

area (page 154)

bar graph (page 89)

brackets (page 112)

calculations (page 2)

capacity (page 142)

Celsius (page 18)

centimetre (page 104)

century (page 128)

chance (page 160)

change (page 20)

check (page 80)

circumference (page 154)

common factor (page 58)

common multiple (page 58)

compare (page 25)

compass (page 157)

composite number (page 60)

cone (page 47)

congruent (page 148)

container (page 142)

convert (page 37)

counting (page 2)

cube (page 48)

cubic (page 142)

cylinder (page 48)

day (page 128)

decade (page 128)

decimal (pages 2, 33, 136)

decimal point (pages 33, 136)

decimetre (page 104)

degree (page 19)

denominator (page 23)

describe (page 12)

diameter (page 154)

difference (page 83)

digit (pages 6, 33, 52)

dimensions (page 152)

divide (pages 110, 124)

divisibility (page 112)

double bar graph (page 91)

duration (page 128)

elapsed (page 131)

equation (page 98)

equivalent (pages 23, 37, 134)

estimate (pages 54, 80)

expression (page 98)

fact (page 75)

fact family (page 77)

factor (page 58)

Fahrenheit (page 18)

fraction (pages 2, 23, 134, 161)

full rotation (page 68)

gallon (page 144)

geometric (page 12)

gram (page 29)

graphs (page 2)

greatest common factor (GCF)
   (page 58)

grouping symbols (page 112)

hundredths (page 37)

inverse (page 110)

irregular polygon (page 42)

kilogram (page 29)

kilometre (page 104)

kite (page 45)

leading figure (page 54)

least common multiple (LCM)
   (page 58)

length (page 104)

line (page 64)

line graph (page 89)

line segment (page 64)

linear (page 104)

litre (page 142)

longer (page 104)

mass (page 29)

mean (page 87)

measurement (page 2)

median (page 87)

metre (page 104)

millilitre (page 142)

millimetre (page 104)

millions (page 6)

minute (page 128)

mixed number (pages 23, 134)

mode (page 87)

multiple (pages 58, 112)

multiply (page 110)

nearest (page 52)

negative (page 19)

non-routine (page 95)

number (pages 2, 6)

numerator (page 23)

numerical (page 12)

obtuse angle (page 68)

occurrence (page 128)

operation (pages 96, 112)

order (page 25)

ounces (page 29)

outcome (page 160)

pair of compasses (page 157)

parallel lines (page 65)

parallelogram (page 44)

part (page 23)